Applied Mathematical Sciences
Volume 83

Applied Mathematical Sciences

1. *John:* Partial Differential Equations, 4th ed.
2. *Sirovich:* Techniques of Asymptotic Analysis.
3. *Hale:* Theory of Functional Differential Equations, 2nd ed.
4. *Percus:* Combinatorial Methods.
5. *von Mises/Friedrichs:* Fluid Dynamics.
6. *Freiberger/Grenander:* A Short Course in Computational Probability and Statistics.
7. *Pipkin:* Lectures on Viscoelasticity Theory.
9. *Friedrichs:* Spectral Theory of Operators in Hilbert Space.
11. *Wolovich:* Linear Multivariable Systems.
12. *Berkovitz:* Optimal Control Theory.
13. *Bluman/Cole:* Similarity Methods for Differential Equations.
14. *Yoshizawa:* Stability Theory and the Existence of Periodic Solution and Almost Periodic Solutions.
15. *Braun:* Differential Equations and Their Applications, 3rd ed.
16. *Lefschetz:* Applications of Algebraic Topology.
17. *Collatz/Wetterling:* Optimization Problems.
18. *Grenander:* Pattern Synthesis: Lectures in Pattern Theory, Vol I.
20. *Driver:* Ordinary and Delay Differential Equations.
21. *Courant/Friedrichs:* Supersonic Flow and Shock Waves.
22. *Rouche/Habets/Laloy:* Stability Theory by Liapunov's Direct Method.
23. *Lamperti:* Stochastic Processes: A Survey of the Mathematical Theory.
24. *Grenander:* Pattern Analysis: Lectures in Pattern Theory, Vol. II.
25. *Davies:* Integral Transforms and Their Applications, 2nd ed.
26. *Kushner/Clark:* Stochastic Approximation Methods for Constrained and Unconstrained Systems
27. *de Boor:* A Practical Guide to Splines.
28. *Keilson:* Markov Chain Models—Rarity and Exponentiality.
29. *de Veubeke:* A Course in Elasticity.
30. *Sniatycki:* Geometric Quantization and Quantum Mechanics.
31. *Reid:* Sturmian Theory for Ordinary Differential Equations.
32. *Meis/Markowitz:* Numerical Solution of Partial Differential Equations.
33. *Grenander:* Regular Structures: Lectures in Pattern Theory, Vol. III.
34. *Kevorkian/Cole:* Perturbation methods in Applied Mathematics.
35. *Carr:* Applications of Centre Manifold Theory.
36. *Bengtsson/Ghil/Källén:* Dynamic Meteorology: Data Assimilation Methods.
37. *Saperstone:* Semidynamical Systems in Infinite Dimensional Spaces.
38. *Lichtenberg/Lieberman:* Regular and Stochastic Motion.
39. *Piccini/Stampacchia/Vidossich:* Ordinary Differential Equations in R^n.
40. *Naylor/Sell:* Linear Operator Theory in Engineering and Science.
41. *Sparrow:* The Lorenz Equations: Bifurcations, Chaos, and Strange Attractors.
42. *Guckenheimer/Holmes:* Nonlinear Oscillations, Dynamical Systems and Bifurcations of Vector Fields.
43. *Ockendon/Tayler:* Inviscid Fluid Flows.
44. *Pazy:* Semigroups of Linear Operators and Applications to Partial Differential Equations.
45. *Glashoff/Gustafson:* Linear Optimization and Approximation: An Introduction to the Theoretical Analysis and Numerical Treatment of Semi-Infinite Programs.
46. *Wilcox:* Scattering Theory for Diffraction Gratings.
47. *Hale et al.:* An Introduction to Infinite Dimensional Dynamical Systems—Geometric Theory.
48. *Murray:* Asymptotic Analysis.
49. *Ladyzhenskaya:* The Boundary-Value Problems of Mathematical Physics.
50. *Wilcox:* Sound Propagation in Stratified Fluids.
51. *Golubitsky/Schaeffer:* Bifurcation and Groups in Bifurcation Theory, Vol. I.
52. *Chipot:* Variational Inequalities and Flow in Porous Media.
53. *Majda:* Compressible Fluid Flow and Systems of Conservation Laws in Several Space Variables.
54. *Wasow:* Linear Turning Point Theory.

(continued following index)

Jerrold Bebernes David Eberly

Mathematical Problems from Combustion Theory

With 14 Illustrations

Springer-Verlag
New York Berlin Heidelberg
London Paris Tokyo Hong Kong

Jerrold Bebernes
Department of Mathematics
University of Colorado
Boulder, CO

David Eberly
Department of Mathematics
University of Texas
San Antonio, TX

AMS Subject Classifications: 35-02, 35Jxx, 35J60, 35Kxx, 35K55

Library of Congress Cataloging-in-Publication Data

Bebernes, Jerrold.
Mathematical problems from combustion theory / Jerrold Bebernes, David Eberly.
p. cm.—(Applied mathematical sciences ; v. 83)
Bibliography: p.
Includes index.
ISBN 0-387-97104-1 (alk. paper)
1. Combustion—Mathematical models. 2. Differential equations, Elliptic. 3. Differential equations, Parabolic. I. Eberly, David. II. Title. III. Series: Applied mathematical science (Springer -Verlag New York Inc.) ; v. 83.
QA1.A647 vol. 83
[QD516]
510 s—dc20
[541.3′61] 89-11483

Photocomposed using the authors' TEX file by the American Mathematical Society.
Printed and bound by R.R. Donnelley & Sons, Harrisonburg, Virginia.
Printed in the United States of America.

9 8 7 6 5 4 3 2 1

ISBN 0-387-97104-1 Springer-Verlag New York Berlin Heidelberg
ISBN 3-540-97104-1 Springer-Verlag Berlin Heidelberg New York

To our wives Sharleen and Shelly

Preface

This monograph evolved over the past five years. It had its origin as a set of lecture notes prepared for the Ninth Summer School of Mathematical Physics held at Ravello, Italy, in 1984 and was further refined in seminars and lectures given primarily at the University of Colorado.

The material presented is the product of a single mathematical question raised by Dave Kassoy over ten years ago. This question and its partial resolution led to a successful, exciting, almost unique interdisciplinary collaborative scientific effort.

The mathematical models described are often times deceptively simple in appearance. But they exhibit a mathematical richness and beauty that belies that simplicity and affirms their physical significance. The mathematical tools required to resolve the various problems raised are diverse, and no systematic attempt is made to give the necessary mathematical background. The unifying theme of the monograph is the set of models themselves.

This monograph would never have come to fruition without the enthusiasm and drive of Dave Eberly–a former student, now collaborator and coauthor–and without several significant breakthroughs in our understanding of the phenomena of blowup or thermal runaway which certain models discussed possess.

A collaborator and former student who has made significant contributions throughout is Alberto Bressan. There are many other collaborators–William Troy, Watson Fulks, Andrew Lacey, Klaus Schmitt–and former students–Paul Talaga and Richard Ely–who must be acknowledged and thanked.

Finally, I would like to acknowledge the continued support of the Army Research Office and its director, Jagdish Chandra.

October 1988

Jerrold Bebernes
University of Colorado

Contents

Preface

1 Introduction **1**
1.1 Basic Fluid Dynamics and Chemical Kinetics 1
1.2 Simplification of the System 6
1.3 Solid Fuel Models 7
1.4 Gaseous Fuel Models 9
1.5 Overview and Comments 13

2 Steady-State Models **15**
2.1 Existence on General Domains 15
2.2 Radial Symmetry 21
2.3 Multiplicity in Special Domains 33
2.4 Solution Profiles 39
2.5 Comments 46

3 The Rigid Ignition Model **47**
3.1 Existence-Uniqueness 48
3.2 Blowup: When? 53
3.3 Blowup: Where? 64
3.4 Blowup: How? 69
3.5 Comments 86

4 The Complete Model for Solid Fuel **88**
4.1 Comparison Techniques 88
4.2 Invariance 94
4.3 Existence 98
4.4 Applications 103
4.5 Comments 106

5 Gaseous Ignition Models **107**
5.1 The Reactive-Diffusive Ignition Model 107
5.2 The Abstract Linear Problem 108
5.3 The Abstract Perturbed Problem 113

5.4 The Radially Symmetric Case 115
5.5 Blowup: Where? 119
5.6 A Nondiffusive Reactive Model 125
5.7 Comments 127

6 Conservation Systems for Reactive Gases **129**
6.1 A Nonreactive Model 129
6.2 Induction Model for a Reactive-Euler System 133
6.3 The Full One-Dimensional Gas Model 136
6.4 Energy Estimates and Density Bounds 140
6.5 Velocity Bounds 152
6.6 Temperature Bounds 157
6.7 Comments 160

References **162**

Index **175**

1

Introduction

1.1 Basic Fluid Dynamics and Chemical Kinetics

Extremely rapid exothermic chemical reactions can develop in combustible materials. For multicomponent reacting mixtures of N chemical species, the complete system of conservation equations can be expressed as

$$\frac{D\vec{w}}{Dt} = \vec{\nabla} \bullet \left(\vec{f}(\vec{w}) \bullet \vec{\nabla}\vec{w}\right) + \vec{g}\left(\vec{w}, \vec{\nabla}\vec{w}\right) \tag{1.1}$$

where $\vec{w}(\vec{x}, t) = (\rho, \vec{u}, T, \vec{y})$ denotes the state of the system and where $\frac{D}{Dt} = \frac{\partial}{\partial t} + \vec{u} \bullet \vec{\nabla}$ is the material derivative. The state $\vec{w}$ includes the density ρ, the temperature T, the mass fractions $\vec{y} = (y_1, \ldots, y_N)$ with $\sum_{i=1}^{N} y_i = 1$, and the mass-average velocity $\vec{u} = \sum_{i=1}^{N} y_i \vec{u}_i$ where $\vec{u}_i$ is the velocity of species i. We also consider the pressure p, which is proportional to density and to temperature. The interaction of the chemistry of the species with the basic fluid flow is described by a highly nonlinear, extremely complex, degenerate, quasilinear parabolic system of partial differential equations. The problem of well-posedness for (1.1) has not been completely resolved [KZH2],[MAT].

Combustible systems composed of gases, liquids, or solids can experience reaction processes which are sustained primarily by a thermal mechanism. This process is typically initiated by boundary heat addition, by localized volumetric heating, by the passage of a dynamic wave, or by very fast compression.

The ignition period process is characterized usually by the appearance of a localized warm region in which the heat production rate accelerates as the reactants are consumed. If conditions are appropriate, the warm region evolves into a region of relatively high temperatures with extremely rapid reaction rates. Subsequent ignition and combustion of the remaining combustible material leads to a significant level of power deposition which is associated with an explosive event.

As a first step in analyzing the combustion process, we will derive the mathematical model from basic conservation principles. The following derivation is essentially that of Williams [WIL1] and Buckmaster and Ludford [BUC1] with a few modifications.

Conservation of Mass. The conservation of mass for each species i is given by

$$\frac{\partial}{\partial t}(\rho y_i) + \vec{\nabla} \bullet (\rho y_i \vec{u}_i) = \dot{r}_i \tag{1.2}$$

where $\dot{r}_i$ is the rate of production (or consumption) for each species i. In a closed system, it is necessary that $\sum_{i=1}^{N} \dot{r}_i = 0$. The equation for the total conservation of mass is obtained by summing the equations in (1.2):

$$\frac{\partial \rho}{\partial t} + \vec{\nabla} \bullet (\rho \vec{u}) = 0. \tag{1.3}$$

Conservation of Species. The equations in (1.2) can be rewritten using (1.3) and the diffusion velocities $\vec{v}_i = \vec{u}_i - \vec{u}$. The resulting equations will be referred to as the conservation of species equations:

$$\rho \left(\frac{\partial y_i}{\partial t} + \vec{u} \bullet \vec{\nabla} y_i \right) = \dot{r}_i - \vec{\nabla} \bullet (\rho y_i \vec{v}_i).$$

We make the assumption that Fick's law applies here: $\rho y_i \vec{v}_i = -\rho D \vec{\nabla} y_i$, where D is the coefficient of diffusion, and is assumed to be the same for all species. Thus, the conservation of species equations are:

$$\rho \left(\frac{\partial y_i}{\partial t} + \vec{u} \bullet \vec{\nabla} y_i \right) = \dot{r}_i + \vec{\nabla} \bullet (\rho D \vec{\nabla} y_i). \tag{1.4}$$

The usual model for the rate of production $\dot{r}_i$ is as follows. For each species i, let N_i be the number of molecules per unit volume and let m_i be the mass of a single molecule. The production rate for a one-step chemical reaction is $\dot{r}_i = m_i \dot{N}_i$. The mass balance is $\sum_{i=1}^{N} \nu_i m_i = \sum_{i=1}^{N} \lambda_i m_i$, where the nonnegative integer values ν_i and λ_i are the stoichiometric coefficients for the reaction. The value ν_i counts the number of molecules of the reactant species i (and is 0 if species i is not a reactant). The value λ_i counts the number of molecules of the product species i (and is 0 if species i is not a product). Consequently,

$$\dot{r}_i = m_i(\lambda_i - \nu_i)\omega \tag{1.5}$$

where $\omega > 0$ measures the rate of reaction. One assumes that ω is proportional to the concentration of each reactant, where species i is counted ν_i times as a reactant. Thus,

$$\omega = A \prod_{j=1}^{N} \left(\frac{\rho y_j}{m_j} \right)^{\nu_j} \tag{1.6}$$

where the proportionality constant A is assumed to be dependent only on temperature. We use the Arrhenius law here and choose

$$A = B_0 T^{\alpha} e^{-E/RT} \tag{1.7}$$

where B_0 and α are constants, E is the activation energy, and

$$R = R_0 \sum_{i=1}^{N} \frac{y_i}{m_i}$$

where R_0 is the universal gas constant. Since T^α reflects a weak temperature dependence, we choose $\alpha = 0$. Therefore, combining (1.5), (1.6), and (1.7) yields

$$\dot{r}_i = m_i(\lambda_i - \nu_i)B_0 e^{-\frac{E}{RT}} \prod_{j=1}^N \left(\frac{\rho y_i}{m_j}\right)^{\nu_j}. \tag{1.8}$$

The conservation of species equations (1.4) become

$$\begin{aligned} \rho\left(\frac{\partial y_i}{\partial t} + \vec{u} \bullet \vec{\nabla} y_i\right) &= \vec{\nabla} \bullet (\rho D \vec{\nabla} y_i) \\ &\quad + m_i(\lambda_i - \nu_i)B_0 e^{-\frac{E}{RT}} \prod_{j=1}^N \left(\frac{\rho y_i}{m_j}\right)^{\nu_j} \end{aligned} \tag{1.9}$$

for $i = 1, \ldots, N$.

Conservation of Momentum. Assuming no momentum is created by the chemical reactions, the conservation of momentum for the total system (treated as a single fluid) is

$$\rho\left(\frac{\partial \vec{u}}{\partial t} + \vec{u} \bullet \vec{\nabla} \vec{u}\right) = \vec{\nabla} \bullet S \tag{1.10}$$

where S is the sum of the stresses in the individual species and the stresses due to the diffusion of species. The interaction of the species produces external forces on each individual species, but the net result is a zero force. We also assume that the gravitational effects are negligible compared to viscous forces.

The stresses for each species are assumed to be of the form

$$S_i = -(p_i + \frac{2}{3}\mu_i \vec{\nabla} \bullet \vec{u})I + 2\mu_i \mathcal{D}_i \tag{1.11}$$

where $\mathcal{D}_i = \frac{1}{2}[\vec{\nabla} \otimes \vec{u}_i + (\vec{\nabla} \otimes \vec{u}_i)^T]$ is the deformation tensor, I is the identity tensor, p_i is pressure, and μ_i is the coefficient of viscosity. Assuming that the intrinsic viscosities of the species are equal, we have $\mu_i = \mu y_i$ for all i. Summing these stresses (and neglecting higher order terms) yields

$$S = -(p + \frac{2}{3}\mu \vec{\nabla} \bullet \vec{u})I + 2\mu \mathcal{D} \tag{1.12}$$

where $\mathcal{D} = \frac{1}{2}[\vec{\nabla} \otimes \vec{u} + (\vec{\nabla} \otimes \vec{u})^T]$. The conservation of momentum equation becomes

$$\rho\left(\frac{\partial \vec{u}}{\partial t} + \vec{u} \bullet \vec{\nabla} \vec{u}\right) = -\vec{\nabla} p + \mu(\Delta \vec{u} + \frac{1}{3}\vec{\nabla}(\vec{\nabla} \bullet \vec{u})). \tag{1.13}$$

Conservation of Energy. Assuming no energy is created by the chemical reactions and the work due to interaction forces is negligible, the conservation of energy for the total system is

$$\rho\left(\frac{\partial \mathcal{E}}{\partial t} + \vec{u} \bullet \vec{\nabla} \mathcal{E}\right) = S : \vec{\nabla} \otimes \vec{u} + \vec{\nabla} \bullet \vec{q} \tag{1.14}$$

where S is the stress tensor, $\vec{q}$ is the sum of the separate energy fluxes and fluxes due to diffusion, and $\mathcal{E}$ is the sum of the separate internal energies. The symbol ":" indicates the dot product of two matrices. Kinetic energies of diffusion are considered to be negligible.

The temperature T is assumed to be the same for all species. The partial pressures are $p_i = \rho R_0 T y_i / m_i$ and so the combined-fluids pressure is

$$p = \sum_{i=1}^{N} \rho R_0 T \frac{y_i}{m_i} = \rho R T. \tag{1.15}$$

The separate internal energies are $\mathcal{E}_i = \mathcal{E}_i(T) = h_i - \frac{p_i}{\rho y_i}$ where the enthalpies h_i are given by

$$h_i = h_i^0 + \int_{T_0}^{T} C_{p_i}(s)\, ds.$$

The value h_i^0 is the heat of formation for species i at standard temperature T_0 and the values C_{p_i} are specific heats at constant pressure. Thus,

$$\mathcal{E} = \sum_{i=1}^{N} h_i y_i - \frac{p}{\rho} = \sum_{i=1}^{N} h_i y_i - RT. \tag{1.16}$$

We assume that the energy flux for each species is due only to heat conduction and the diffusion of species, so the flux for species i is

$$\vec{q_i} = k_i \vec{\nabla} T - \rho y_i h_i \vec{v_i} \tag{1.17}$$

where k_i is the coefficient of thermal conductivity. Using Fick's law ($\rho y_i \vec{v_i} = -\rho D \vec{\nabla} y_i$) and summing the equations in (1.17) yields

$$\vec{q} = k \vec{\nabla} T - \sum_{i=1}^{N} \rho D h_i \vec{\nabla} y_i \tag{1.18}$$

where $k = \sum_{i=1}^{N} k_i$.

We wish to convert the energy conservation equation (1.14) into one involving the temperature T. Define

$$C_p = \sum_{i=1}^{N} y_i C_{p_i}(T) \text{ and } C_v = \sum_{i=1}^{N} y_i C_{v_i}(T)$$

where C_{v_i} is specific heat at constant volume. Since $C_{p_i} - C_{v_i} = R_0/m_i$, we have

$$C_p - C_v = R. \tag{1.19}$$

We assume that $C_p = C_p(T)$ and $C_v = C_v(T)$. As a consequence,

$$\sum_{i=1}^{N} \vec{\nabla} y_i \bullet \vec{\nabla} h_i = \sum_{i=1}^{N} C_{p_i} \vec{\nabla} y_i \bullet \vec{\nabla} T = \vec{\nabla} C_p \bullet \vec{\nabla} T = 0$$

and

$$\begin{aligned} \vec{\nabla} \bullet \vec{q} &= \vec{\nabla} \bullet (k \vec{\nabla} T) + \textstyle\sum_{i=1}^{N} \vec{\nabla} \bullet (\rho D h_i \vec{\nabla} y_i) \\ &= \vec{\nabla} \bullet (k \vec{\nabla} T) + \textstyle\sum_{i=1}^{N} h_i \vec{\nabla} \bullet (\rho D \vec{\nabla} y_i). \end{aligned}$$

The left-hand side of the energy conservation equation is

$$\begin{aligned} \rho\left(\tfrac{\partial \mathcal{E}}{\partial t} + \vec{u} \bullet \vec{\nabla} \mathcal{E}\right) &= \textstyle\sum_{i=1}^{N} h_i \rho \left(\tfrac{\partial y_i}{\partial t} + \vec{u} \bullet \vec{\nabla} y_i\right) \\ &\quad + \textstyle\sum_{i=1}^{N} \rho y_i \left(\tfrac{\partial h_i}{\partial t} + \vec{u} \bullet \vec{\nabla} h_i\right) - \rho R \left(\tfrac{\partial T}{\partial t} + \vec{u} \bullet \vec{\nabla} T\right) \\ &= \textstyle\sum_{i=1}^{N} h_i [\vec{\nabla} \bullet (\rho D \vec{\nabla} y_i) + \dot{r}_i] + \rho C_p \left(\tfrac{\partial T}{\partial t} + \vec{u} \bullet \vec{\nabla} T\right) \\ &\quad - \rho R \left(\tfrac{\partial T}{\partial t} + \vec{u} \bullet \vec{\nabla} T\right) \\ &= \textstyle\sum_{i=1}^{N} h_i [\vec{\nabla} \bullet (\rho D \vec{\nabla} y_i) + \dot{r}_i] + \rho C_v \left(\tfrac{\partial T}{\partial t} + \vec{u} \bullet \vec{\nabla} T\right) \end{aligned}$$

where we have used the species conservation equation (1.4) and the identity (1.19). The right-hand side of the energy equation is

$$\begin{aligned} S : \vec{\nabla} \otimes \vec{u} + \vec{\nabla} \bullet \vec{q} &= -(p + \tfrac{2}{3} \mu \vec{\nabla} \bullet \vec{u})(\vec{\nabla} \bullet \vec{u}) + 2\mu(\mathcal{D} : \vec{\nabla} \otimes \vec{u}) \\ &\quad + \vec{\nabla} \bullet (k \vec{\nabla} T) + \textstyle\sum_{i=1}^{N} h_i \vec{\nabla} \bullet (\rho D \vec{\nabla} y_i). \end{aligned}$$

Equating the two sides yields

$$\begin{aligned} &\rho C_v \left(\tfrac{\partial T}{\partial t} + \vec{u} \bullet \vec{\nabla} T\right) = \\ &\quad \vec{\nabla} \bullet (k \vec{\nabla} T) - p(\vec{\nabla} \bullet \vec{u}) + 2\mu[-\tfrac{1}{3}(\vec{\nabla} \bullet \vec{u})^2 + \mathcal{D} : \vec{\nabla} \otimes \vec{u}] \\ &\quad - \textstyle\sum_{i=1}^{N} h_i m_i (\lambda_i - \nu_i) B_0 e^{-\frac{E}{RT}} \prod_{j=1}^{N} \left(\tfrac{\rho y_j}{m_j}\right)^{\nu_j}. \end{aligned} \tag{1.20}$$

The complete system of conservation laws for combustion can be summarized from (1.3), (1.9), (1.13), and (1.20) as

$$\begin{aligned} &\tfrac{\partial \rho}{\partial t} + \vec{\nabla} \bullet (\rho \vec{u}) = 0 \\ &\rho\left(\tfrac{\partial \vec{u}}{\partial t} + \vec{u} \bullet \vec{\nabla} \vec{u}\right) = -\vec{\nabla} p + \mu(\Delta \vec{u} + \tfrac{1}{3} \vec{\nabla}(\vec{\nabla} \bullet \vec{u})) \\ &\rho C_v \left(\tfrac{\partial T}{\partial t} + \vec{u} \bullet \vec{\nabla} T\right) = \\ &\quad \vec{\nabla} \bullet (k \vec{\nabla} T) - p(\vec{\nabla} \bullet \vec{u}) + 2\mu[-\tfrac{1}{3}(\vec{\nabla} \bullet \vec{u})^2 + \mathcal{D} : \vec{\nabla} \otimes \vec{u}] \\ &\quad - \textstyle\sum_{i=1}^{N} h_i m_i (\lambda_i - \nu_i) B_0 e^{-\frac{E}{RT}} \prod_{j=1}^{N} \left(\tfrac{\rho y_j}{m_j}\right)^{\nu_j} \\ &\rho\left(\tfrac{\partial y_i}{\partial t} + \vec{u} \bullet \vec{\nabla} y_i\right) = \\ &\quad \vec{\nabla} \bullet (\rho D \vec{\nabla} y_i) + m_i (\lambda_i - \nu_i) B_0 e^{-\frac{E}{RT}} \prod_{j=1}^{N} \left(\tfrac{\rho y_j}{m_j}\right)^{\nu_j} \\ &p = \rho R T. \end{aligned} \tag{1.21}$$

1.2 Simplification of the System

The single one-step irreversible reaction that we will consider is of the form:

$$\nu_F F + \nu_O O \to \lambda_P P,$$

where F represents fuel, O represents oxidant, P represents the product, and where ν_F, ν_O, and λ_P are stoichiometric constants. This reaction involves three mass fractions: y_F, y_O, and y_P. If both fuel and oxidant are present in the correct proportions, then both are entirely consumed in the process. In this case the initial values y_{F_0} and y_{O_0} are of the same order of magnitude, so the reaction rate is strongly dependent on both mass fractions. However, if $y_{F_0} \gg y_{O_0}$, then the reaction rate is weakly dependent on y_F since y_F does not change much. Since y_F is approximately constant, we ignore its species equation and consider only the single species equation for y_O.

Note that the stoichiometric mixture of fuel and oxidant satisfies $\frac{y_O}{y_F} \doteq \frac{\nu_O}{\nu_F}$. Choose $m = \nu_F + \nu_O$, $\lambda_O = 0$, $h = h_O$, $y = y_O$, and

$$B = B_0 \nu_O m_O \left(\frac{\nu_F}{\nu_O}\right)^{\nu_F} m_O^{-\nu_O} m_F^{-\nu_F};$$

then (1.21) becomes

$$\begin{aligned}
&\rho_t + \vec{\nabla} \bullet (\rho\vec{u}) = 0 \\
&\rho(\vec{u}_t + \vec{u} \bullet \vec{\nabla}\vec{u}) = -\vec{\nabla}p + \mu[\Delta\vec{u} + \tfrac{1}{3}\vec{\nabla}(\vec{\nabla} \bullet \vec{u})] \\
&\rho C_v(T_t + \vec{u} \bullet \vec{\nabla}T) = \\
&\quad \vec{\nabla} \bullet (k\vec{\nabla}T) - p(\vec{\nabla} \bullet \vec{u}) \\
&\quad + 2\mu[D : \vec{\nabla} \otimes \vec{u} - \tfrac{1}{3}(\vec{\nabla} \bullet \vec{u})^2] + Bh\rho^m y^m e^{-\frac{E}{RT}} \\
&\rho(y_t + \vec{u} \bullet \vec{\nabla}y) = \vec{\nabla} \bullet (\rho D\vec{\nabla}y) - B\rho^m y^m e^{-\frac{E}{RT}} \\
&p = \rho RT.
\end{aligned} \tag{1.22}$$

The combustion model (1.22) can be nondimensionalized in a rational manner in order to elucidate the significant parameters. Assume initially that a reactive, viscous, heat conducting, compressible gas is in an equilibrium state defined by the dimensional quantities $p_0 = p(\vec{x}, 0)$, $\rho_0 = \rho(\vec{x}, 0)$, $T_0 = T(\vec{x}, 0)$, $y_0 = y(\vec{x}, 0)$, and $\vec{u}_0 = \vec{u}(\vec{x}, 0)$.

At time $t = 0$, a small initial disturbance is created on a length scale L. Define $\underline{\vec{x}} = \vec{x}/L$ as the new position vector. Let t_R be a reference time (to be determined later). Define $\overline{t} = t/t_R$ as the new time scale. Nondimensionalize the system variables: $\overline{p} = p/p_0$, $\overline{\rho} = \rho/\rho_0$, $\overline{T} = T/T_0$, $\overline{y} = y/y_0$, and $\underline{\vec{u}} = \vec{u}/(L/t_R)$. Also nondimensionalize the quantities: $\overline{\mu} = \mu/\mu_0$, $\overline{D} =$

D/D_0, $\overline{C}_p = C_p/C_{p_0}$, $\overline{C}_v = C_v/C_{v_0}$, $\overline{k} = k/k_0$, and $\overline{K} = K/K_0$, where $K = k/(\rho C_p)$ is thermal diffusivity.

In the scaling of the system we will use the following quantities: $\gamma = C_{p_0}/C_{v_0}$, the gas parameter; $\varepsilon = RT_0/E$, the nondimensional inverse of the activation energy; $Pr = C_{p_0}\mu_0/k_0$, the Prandtl number; $Le = D_0/K_0$, the Lewis number; $C_0 = \sqrt{\gamma R T_0}$, the initial sound speed; $t_A = L/C_0$, the acoustic time scale; $t_C = L^2/K_0$, the conduction time scale; and $\overline{h} = hy_0/(C_{v_0}T_0)$, the nondimensional heat of reaction.

Substituting these into (1.22) and dropping the bar notation gives us the nondimensional model

$$
\begin{aligned}
&\rho_t + \vec{\nabla} \bullet (\rho\vec{u}) = 0 \\
&\rho(\vec{u}_t + \vec{u} \bullet \vec{\nabla}\vec{u}) = -\tfrac{1}{\gamma}\left(\tfrac{t_R}{t_A}\right)^2 \vec{\nabla}p + Pr\left(\tfrac{t_R}{t_C}\right)\mu[\Delta\vec{u} + \tfrac{1}{3}\vec{\nabla}(\vec{\nabla} \bullet \vec{u})] \\
&\rho C_v(T_t + \vec{u} \bullet \vec{\nabla}T) = \\
&\quad \gamma\left(\tfrac{t_R}{t_C}\right)\vec{\nabla} \bullet (k\vec{\nabla}T) - (\gamma - 1)p(\vec{\nabla} \bullet \vec{u}) \\
&\quad + 2\mu\gamma(\gamma-1)Pr\left(\tfrac{t_A^2}{t_R t_C}\right)[\mathcal{D} : \vec{\nabla} \otimes \vec{u} - \tfrac{1}{3}(\vec{\nabla} \bullet \vec{u})^2] \\
&\quad + t_R B h \rho^m y^m e^{-\frac{1}{\varepsilon T}} \\
&\rho(y_t + \vec{u} \bullet \vec{\nabla}y) = Le\left(\tfrac{t_R}{t_C}\right)\vec{\nabla} \bullet (\rho D\vec{\nabla}y) - t_R B\rho^m y^m e^{-\frac{1}{\varepsilon T}} \\
&p = \rho T.
\end{aligned}
\qquad (1.23)
$$

In addition to the simplifying single-species chemistry assumption, we will use the method of activation energy asymptotics to obtain simpler models of the combustion process. In (1.23c,d), the reaction terms contain a term of the form $\exp(-\frac{1}{\varepsilon T})$. Activation energy asymptotics is concerned with the asymptotic expansion of solutions as $\varepsilon \to 0^+$. Usually different scalings of independent variables are involved. We will always work with the scaling $\vec{x} \mapsto \vec{x}/L$, but the reference time t_R and scaling $t \mapsto t/t_R$ will be chosen to select those aspects of the model that we are interested in.

In the following sections we will develop various models based on activation energy asymptotics, which we refer to as the *ignition* or *induction period* models.

1.3 Solid Fuel Models

The traditional theory of thermal reaction processes is formulated for nondeformable materials of constant density. Conceptually, this system is much simpler than for compressible gases. If the single chemical species is a solid in a bounded container $\Omega \subset \mathbb{R}^3$, then $\vec{u} = 0$, $\rho = 1$, $\gamma = 1$, and the ratio $t_R/t_C = O(1)$. Thus, (1.23) reduces to the reaction-diffusion system which can be written as

$$\begin{aligned} T_t - \Delta T &= \varepsilon \delta y^m \exp(\tfrac{T-1}{\varepsilon T}) \\ y_t - \beta \Delta y &= -\varepsilon \delta \Gamma y^m \exp(\tfrac{T-1}{\varepsilon T}) \end{aligned}, \quad (x,t) \in \Omega \times (0,\infty) \tag{1.24}$$

with initial-boundary conditions

$$\begin{aligned} &T(x,0) = 1,\ y(x,0) = 1,\quad x \in \Omega \\ &T(x,t) = 1,\ \tfrac{\partial y(x,t)}{\partial \eta(x)} = 0,\quad (x,t) \in \partial\Omega \times (0,\infty) \end{aligned} \tag{1.25}$$

where $\beta \geq 0$, $\Gamma > 0$, and $\delta > 0$ is the Frank-Kamenetski parameter.

Until relatively recently, even this system was considered intractable and was approximated by simpler models. One method of simplification is to identify and restrict the range of certain parameters and then use an asymptotic analysis. For all fuels of interest, the parameter ε is assumed small. By using the method of activation energy asymptotics, and letting $T = 1 + \varepsilon\theta$ and $y = 1 - \varepsilon c$ be the first order approximations, IBVP (1.24)-(1.25) can be written as

$$\begin{aligned} \theta_t - \Delta\theta &= \delta(1-\varepsilon c)^m \exp(\tfrac{\theta}{1+\varepsilon\theta}) \\ c_t - \beta\Delta c &= \delta\Gamma(1-\varepsilon c)^m \exp(\tfrac{\theta}{1+\varepsilon\theta}) \end{aligned}, \quad (x,t) \in \Omega \times (0,\infty) \tag{1.26}$$

with initial-boundary conditions

$$\begin{aligned} &\theta(x,0) = 0,\ c(x,0) = 0,\quad x \in \Omega \\ &\theta(x,t) = 0,\ \tfrac{\partial c(x,t)}{\partial \eta(x)} = 0,\quad (x,t) \in \partial\Omega \times (0,\infty). \end{aligned} \tag{1.27}$$

For $\varepsilon \ll 1$, the activation energy method has essentially decoupled (1.26) and we need only consider the *solid fuel ignition model*

$$\theta_t - \Delta\theta = \delta e^{\theta},\quad (x,t) \in \Omega \times (0,\infty) \tag{1.28}$$

with initial-boundary conditions

$$\begin{aligned} &\theta(x,0) = 0,\quad x \in \Omega \\ &\theta(x,t) = 0,\quad (x,t) \in \partial\Omega \times (0,\infty) \end{aligned} \tag{1.29}$$

and the associated *steady-state model*

$$-\Delta\psi = \delta e^{\psi},\quad x \in \Omega \tag{1.30}$$

$$\psi(x) = 0,\quad x \in \partial\Omega. \tag{1.31}$$

We also will consider the *small fuel loss steady-state model*

$$-\Delta\phi = \delta \exp\left(\frac{\phi}{1+\varepsilon\phi}\right),\quad x \in \Omega \tag{1.32}$$

$$\phi(x) = 0,\quad x \in \partial\Omega. \tag{1.33}$$

The thermal reaction process in a rigid material during the ignition period is modeled by the solid fuel ignition model (1.28)-(1.29). Its solution should predict the time-history of the spatially-varying reaction process. This process depends only on a pointwise balance between chemically generated heat addition and heat transfer by conduction. If the heat loss is sufficiently large compared with the energy release associated with the strongly temperature dependent reaction rate, then energy equilibrium is established. In this case the chemical reaction time is commensurate with the container time scale for conduction. The maximum system temperature is never much different from the initial value because so much energy is lost to the relatively cold boundary. In this type of reaction, the reactant species is eventually consumed.

In contrast, when the heat loss is sufficiently small, a localized temperature rise occuring at first on the conduction time scale will cause the reaction rate to accelerate dramatically. As a result, the characteristic chemical time becomes much shorter than the conduction time during the induction period. When that occurs, a sharply focused temperature region appears in which the fuel is rapidly depleted. The explosive burst of power generation provides an essential distinction between a benign subcritical event and this more dynamic supercritical process.

For a solid reactive fuel in a bounded container, the associated thermal event can be either violent or mild in the sense described above. If the thermal event is violent, then it is said to be *supercritical* or *explosive.* If it is not, then the event is said to be *subcritical* or a *fizzle.*

Of these solid fuel models, we will be primarily concerned with IBVP (1.28)- (1.29) and BVP (1.30)-(1.31), and generalizations thereof. Detailed information on these models can be found in Chapters 2 and 3. Qualitative properties for the complete solid fuel model (1.26)-(1.27) can be found in Chapter 4.

1.4 Gaseous Fuel Models

If the chemical species is a warm reactive compressible ideal gas embedded in an infinite field of a cooler reactive or inert gas, or contained in a bounded container Ω, then the complete model (1.23) does not immediately simplify as for a solid fuel and the problem of well-posedness for (1.23) is unresolved.

We thus develop an induction period theory for a system with a high activation energy reaction. The character of the induction period models depends intimately on the ratios formed from the characteristic chemical time, the acoustic time, and the conduction time of the embedded warm region. A systematic investigation of the different ratios permits one to predict the type of thermal explostion to be expected for a given physicochemical system.

Assume a high activation energy reaction ($\varepsilon \ll 1$) and set

$$\rho = 1 + \varepsilon M, \quad p = 1 + \varepsilon P, \quad T = 1 + \varepsilon\theta, \quad \vec{u} = \varepsilon\vec{v}, \quad y = 1 - \varepsilon c, \tag{1.34}$$

assuming that the initial temperature disturbance remains small. If $O(\varepsilon)$ terms are ignored, we obtain the induction model for a gaseous system using (1.34):

$$\begin{aligned}
&M_t + \vec{\nabla} \bullet \vec{v} = 0\\
&\vec{v}_t = -\tfrac{1}{\gamma}\left(\tfrac{t_R}{t_A}\right)^2 \vec{\nabla}P + Pr\left(\tfrac{t_R}{t_C}\right)\mu[\Delta\vec{v} + \tfrac{1}{3}\vec{\nabla}(\vec{\nabla} \bullet \vec{v})]\\
&\theta_t = t_R Bh\varepsilon^{-1}e^{-\frac{1}{\varepsilon}}e^{\theta} + \gamma\left(\tfrac{t_R}{t_C}\right)\Delta\theta - (\gamma - 1)\vec{\nabla} \bullet \vec{v}\\
&\quad + 2\gamma(\gamma - 1)\tfrac{t_A^2}{t_R t_C}\mu\varepsilon Pr[-\tfrac{1}{3}(\vec{\nabla} \bullet \vec{v})^2\\
&\quad + \{\vec{\nabla} \otimes \vec{v} + (\vec{\nabla} \otimes \vec{v})^T\} : \vec{\nabla} \otimes \vec{v}]\\
&c_t = t_R B\varepsilon^{-1}e^{-\frac{1}{\varepsilon}}e^{\theta} + Le\left(\tfrac{t_R}{t_C}\right)\Delta c\\
&P = M + \theta.
\end{aligned} \tag{1.35}$$

The induction model (1.35) contains three time scales: t_R, t_A, and t_C, which depend on the given thermochemical system with the reference time t_R yet to be specified. If we assume that the perturbation temperature θ and the concentration c variations are caused by the chemical reaction process, then for ε small these should be a balance of the accumulation terms θ_t and c_t in (1.35) with the reaction terms involving e^{θ}. We therefore define the reference time

$$t_r = \frac{\varepsilon e^{1/\varepsilon}}{B} \tag{1.36}$$

which represents the chemical time multiplied by ε. The reduced ignition models depend directly on the ratios of these time scales.

If the chemical and conduction times are of the same duration so that

$$a := \frac{t_R}{t_A} = O(1),$$

then the induction momentum, energy, and species equations of (1.35) can be written as

$$\begin{aligned}
&\left(\tfrac{t_A}{t_C}\right)^2 \vec{v}_t = -\tfrac{a^2}{\gamma}\vec{\nabla}P + a\left(\tfrac{t_A}{t_C}\right)^2 \mu Pr[\Delta\vec{v} + \tfrac{1}{3}\vec{\nabla}(\vec{\nabla} \bullet \vec{v})]\\
&\theta_t = he^{\theta} + a\gamma\Delta\theta - (\gamma - 1)\vec{\nabla} \bullet \vec{v}\\
&c_t = e^{\theta} + a\,Le\Delta c.
\end{aligned} \tag{1.37}$$

For spatially macroscopic initial disturbances, we may assume that $t_A/t_C \ll 1$. From the inductive momentum equation in (1.37), we see that to a first approximation $P = P(t)$. Thus, from the mass equation in (1.35) and the energy equation in (1.37) we have

$$\theta_t = \frac{h}{\gamma} e^{\theta} + a\Delta\theta + \frac{\gamma - 1}{\gamma} P'(t). \tag{1.38}$$

For a bounded container Ω, since the total mass must be conserved,

$$\int_{\Omega} \rho(x,t)\, dx = \text{vol}(\Omega),$$

which implies $\int_{\Omega} M(x,t)\, dx = 0$. Thus,

$$\int_{\Omega} P(t)\, dx = \int_{\Omega} \theta(x,t)\, dx$$

from $P = M + \theta$ and hence

$$P(t) = \frac{1}{\text{vol}(\Omega)} \int_{\Omega} \theta(x,t)\, dx.$$

We can thus rewrite (1.38) as

$$\theta_t - a\Delta\theta = \delta e^{\theta} + \frac{\gamma - 1}{\gamma} \frac{1}{\text{vol}(\Omega)} \int_{\Omega} \theta_t(x,t)\, dx \tag{1.39}$$

and impose initial-boundary conditions of the type

$$\begin{aligned} &\theta(x,0) = \theta_0(x), \quad x \in \Omega \\ &\theta(x,t) = 0, \quad (x,t) \in \partial\Omega \times (0,\infty). \end{aligned} \tag{1.40}$$

This particular model will be referred to as the *gaseous reactive-diffusive ignition model.* Note that for $\gamma = 1$ this model reduces to the classical solid fuel ignition model. The model (1.39)-(1.40) is analyzed in Chapter 5.

If the ratio $a = t_R/t_C \ll O(1)$, so that the chemical time is much shorter than the conduction time in the domain Ω, then three subcases arise, all of which exhibit nondiffusive phenomena.

If $t_A \ll t_R \ll t_C$, then again $P = P(t)$ is spatially independent and the energy equation becomes

$$\theta_t = he^{\theta} - (\gamma - 1)\vec{\nabla} \bullet \vec{v}$$

which can be rewritten as

$$\theta_t = \frac{h}{\gamma} e^{\theta} + \frac{\gamma - 1}{\gamma} \frac{1}{\text{vol}(\Omega)} \int_{\Omega} \theta_t(x,t)\, dx. \tag{1.41}$$

With initial condition (1.40a), this is a nondiffusive version of (1.39)-(1.40) and will be treated briefly in Chapter 5.

If $O(t_R) = t_A \ll t_C$, then (1.35) reduces to

$$\begin{aligned} \theta_t - \tfrac{\gamma-1}{\gamma} P_t &= \tfrac{h}{\gamma} e^\theta \\ \vec{v}_t + \tfrac{a^2}{\gamma}\left(\tfrac{t_C}{t_A}\right)^2 \vec{\nabla} P &= 0 \\ \vec{\nabla} \bullet \vec{v} + \tfrac{1}{\gamma} P_t &= \tfrac{h}{\gamma} e^\theta \end{aligned} \tag{1.42}$$

where $P = P(x,t)$. This reactive-Euler model will be discussed in Chapter 6.

In case $t_R \ll t_A \ll t_C$, the velocity perturbation $\vec{v} = \vec{v}(x)$ is independent of time which implies dominant inertial confinement of the heated fluid.

An induction period theory for a reactive perfect gas is modeled for the various time ratios by (1.39), (1.41), and (1.42). The evolution of these systems depends on the relative effects of the chemical power deposition, conductive heat transfer, energy convection, and compressive heating. Once again the associated thermal event can be either violent or mild in the sense desribed earlier.

Finally, in Chapter 6 we analyze the full gaseous model given by (1.23). Gradient systems and conservation laws are discussed first. We then proceed to analyze (1.23) in the special case of a heat-conductive viscous reactive compressible gas bounded by two parallel plates. In Euler coordinates the model is given by

$$\begin{aligned} &\rho_t + (v\rho)_y = 0 \\ &\rho[v_t + vv_y] = \lambda_1 v_{yy} - k(\rho\theta)_y \\ &\rho[\theta_t + v\theta_y] = \lambda_2 \theta_{yy} + \lambda_1 v_y^2 - k\rho\theta v_y + \delta\rho f(\rho,\theta,z) \\ &\rho[z_t + vz_y] = \lambda_3(\rho z_y)_y - \rho f(\rho,\theta,z) \end{aligned} \tag{1.43}$$

where k, δ, and $\lambda_i \ \ (i = 1,2,3)$ are positive constants, where $t \geq 0$ is the time, and where $y \in [0,1] \subset \mathbb{R}$ is the one -dimensional space variable. The variables ρ, v, θ, and z represent the density, velocity, temperature, and concentration of unburned fuel, respectively. Let $\Omega = (0,1)$ and $\partial\Omega = \{0,1\}$. The initial conditions for (1.43) will be

$$\begin{aligned} \rho(y,0) = \rho_0(y), \ \ v(y,0) = v_0(y) \\ \theta(y,0) = \theta_0(y), \ \ z(y,0) = z_0(y) \end{aligned} \quad , \ y \in \Omega. \tag{1.44}$$

For a thermally insulated boundary, the boundary conditions are

$$\begin{aligned} &v(y,t) = 0, \ \ \theta_y(y,t) = 0 \\ &z_y(y,t) = 0 \end{aligned} \quad , \ (y,t) \in \partial\Omega \times (0,\infty). \tag{1.45}$$

For a noninsulated boundary, the boundary conditions are

$$\left.\begin{aligned} &v(y,t)=0, \quad z_y(y,t)=0 \\ &a[\theta(y,t)-T]-b\theta_y(y,t)=0 \end{aligned}\right., \ (y,t)\in\partial\Omega\times(0,\infty), \tag{1.46}$$

where $a \geq 0$, $b \geq 0$, $a+b>0$, and $T>0$. Under the appropriate assumptions, global existence and uniqueness are proved for this model.

1.5 Overview and Comments

The spatially-varying transient process describing a thermal event should be entirely predictable for a given set of physical properties, system geometry, and initial-boundary conditions. For the various initial-boundary value problems which model a reactive thermal event, the following questions naturally arise:

1. Do these models give a reasonable time-history description of the state of the system?

2. Do the various models distinguish between explosive and nonexplosive events?

3. If the thermal event is explosive, can one predict precisely when the thermal explosions will occur, determine where the hotspots will develop, and finally predict how the hotspot or blowup singularities evolve as the blowup time is approached?

4. How do the various models compare?

In the next five chapters we will address these questions. Since we are primarily interested in explosive thermal events, we will extensively answer question number three, and refer to the three aspects of this problem as: *Blowup - When, Where, and How.*

The brief derivation of the governing conservation equations (1.22) of combustion presented in this chapter follows the treatment given in the books by Williams [WIL1], Buckmaster and Ludford [BUC1], and Strehlow [STR]. The nondimensionalization of (1.22) for a single species to system (1.23) is based on the work of Kassoy, Kapila, and Stewart [KAP2].

The classical complete model (1.24)-(1.25) for a solid fuel has a theory which is now relatively complete. The basic existence-uniqueness results were proved independently by Bebernes, Chueh, and Fulks [BEB3] and by Amann [AMA4] using invariance techniques.

The idea of using activation energy asymptotics to derive the ignition models from the complete system (1.24) can be traced back to the pioneering work of Frank-Kamenetski [FRA]. The idea was put on firmer ground by Williams [WIL2] in 1971 and has been systematically carried forward by a number of combustion theorists such as Buckmaster, Ludford, Kapila, Kassoy, Liñan, and their co-workers.

Much remains to be done to get a reasonable mathematical theory developed for (1.23) in higher space dimensions, although quite a bit is known for the one-dimensional model (1.43). This will be discussed in Chapter 6. The ignition model (1.35) has reduced forms depending on the ratio of time scales which are amenable to a rigorous mathematical treatment.

The gaseous reactive-diffusive ignition model (1.39)-(1.40) was first derived by Kassoy and Poland [KAS5] in 1983 and was further considered by Bebernes and Bressan [BEB5]. The reactive-Euler ignition model (1.41) is relatively easily analyzed and was done so in [BEB13]. Model (1.42) has been considered by Kapila, Jackson, and Stewart [JAC1], [JAC2], and by Majda [MAJ6].

2

Steady-State Models

The first section of this chapter deals with existence for the Dirichlet problem where the nonlinearity $F(x, u)$ is a nonnegative function. The key result used is an existence theorem based on *a priori* knowledge of upper and lower solutions. We also analyze the spectrum of nonlinear eigenvalue problems and determine bounds on the critical eigenvalues. As applications we consider the Gelfand problem where $f(u) = \exp(u)$, and we consider the perturbed Gelfand problem where $f(u) = \exp(\frac{u}{1+\varepsilon u})$.

The second section deals with a powerful result by Gidas, Ni, and Nirenberg. For a nonlinearity $f \in C^1$ in the Dirichlet problem on a ball in $\mathbb{R}^n$, any positive solution is radially symmetric and radially decreasing. The proof uses maximum principles and the method of moving parallel planes. We also discuss the overdetermined Dirichlet problem where $u = 0$ and $\frac{\partial u}{\partial \eta} = c$ on the boundary of a set Ω. We prove that Ω is necessarily a ball in $\mathbb{R}^n$.

In the third section we give the proof of multiplicity for the Gelfand problem $[f(u) = \exp(u)]$ on a ball in $\mathbb{R}^n$. The theorem is due to Joseph and Lundgren. Similar multiplicity results were developed by Dancer for the perturbed Gelfand problem $[f(u) = \exp(\frac{u}{1+\varepsilon u})]$, but the approach is significantly different. We state the theorem without proof and illustrate it with bifurcation diagrams.

Finally, the fourth section deals with the qualitative shape of solutions to both the Gelfand and the perturbed Gelfand problem. The proofs involve a detailed analysis of the bifurcation diagrams that accompany the problems. These results are due to Bebernes, Eberly, and Fulks.

2.1 Existence on General Domains

Consider the boundary value problem

$$\begin{aligned} -\Delta u &= f(x, u), \quad x \in \Omega \\ u(x) &= \theta(x), \quad x \in \partial\Omega \end{aligned} \tag{2.1}$$

where $\Omega \subset \mathbb{R}^n$ is a bounded domain whose boundary $\partial\Omega$ is an $(n-1)$-dimensional manifold of class $C^{2+\alpha}$ for some $\alpha \in (0, 1)$. That is, for every $x = (x_1, \ldots, x_n) \in \partial\Omega$ there exists aneighborhood N of x such that $\partial\Omega \cap N$ may be represented as $x_i = h(x_1, \ldots, x_{i-1}, x_{i+1}, \ldots, x_n)$ for some i where $h \in C^{2+\alpha}(\mathbb{R}^{n-1}, \mathbb{R})$. Assume $f \in C^\alpha(\overline{\Omega} \times \mathbb{R}, \mathbb{R})$ and $\theta \in C(\partial\Omega, \mathbb{R})$. A solution of BVP (2.1) is a function $u \in C(\overline{\Omega}, \mathbb{R}) \cap C^2(\Omega, \mathbb{R})$ which satisfies both equations in (2.1).

Definition 2.1 *A function $\alpha \in C(\overline{\Omega}, \mathbb{R}) \cap C^2(\Omega, \mathbb{R})$ is a* lower solution *of BVP(2.1) if*

$$-\Delta\alpha(x) \leq f(x, \alpha(x)), \quad x \in \Omega$$
$$\alpha(x) \leq \theta(x), \quad x \in \partial\Omega.$$

An upper solution *$\beta(x)$ is defined similarly where the inequalities above are reversed.*

Theorem 2.1 *If BVP (2.1) has a lower solution $\alpha(x)$ and an upper solution $\beta(x)$ with $\alpha(x) \leq \beta(x)$ on $\overline{\Omega}$, then BVP (2.1) has a solution $u(x)$ with $\alpha(x) \leq u(x) \leq \beta(x)$ on $\overline{\Omega}$.*

The proof of Theorem 2.1 can be found in Schmitt [SCH] and uses degree theoretic methods.

Consider the nonlinear eigenvalue problem

$$\begin{aligned} -\Delta u &= \lambda F(x, u), \quad x \in \Omega \\ u(x) &= 0, \quad x \in \partial\Omega \end{aligned} \tag{2.2}$$

assuming $F \in C^{\alpha}(\overline{\Omega} \times \mathbb{R}, [0, \infty))$ and $\lambda \in \mathbb{R}$.

Lemma 2.2 *For $\lambda \geq 0$, $\alpha(x) \equiv 0$ is a lower solution for BVP (2.2).*

Proof. Observe that $\alpha(x)$ satisfies $-\Delta\alpha(x) = 0 \leq \lambda F(x, 0) = \lambda F(x, \alpha(x))$ for $\lambda \geq 0$ since $F(x, u) \geq 0$. □

Definition 2.2 *The* spectrum Σ *of BVP (2.2) is the set of all $\lambda \in \mathbb{R}$ such that (2.2) has a nonnegative solution.*

Lemma 2.3 *If $\lambda_1 \in \Sigma \cap (0, \infty)$, then $[0, \lambda_1] \subset \Sigma$.*

Proof. By Lemma 2.2, $\alpha(x) \equiv 0$ is a lower solution for (2.2). Let $\beta(x)$ be a nonnegative solution of (2.2) with $\lambda = \lambda_1$; then

$$-\Delta\beta(x) = \lambda_1 F(x, \beta(x)) \geq \lambda F(x, \beta(x))$$

with $\beta(x) = 0$ on $\partial\Omega$. Thus, $\beta(x)$ is an upper solution of (2.2) for any $\lambda \in [0, \lambda_1)$ with $\beta(x) \geq \alpha(x)$ on $\overline{\Omega}$. By Theorem 2.1, (2.2) has a nonnegative solution for any $\lambda \in [0, \lambda_1]$. □

Lemma 2.4 *Assume there exist functions* $f_0, r \in C^\alpha(\overline{\Omega}, (0,\infty))$ *such that*

$$F(x,u) \geq f_0(x) + r(x)u, \quad (x,u) \in \overline{\Omega} \times [0,\infty);$$

then BVP (2.2) has no nonnegative solutions for $\lambda \geq \lambda_0(r)$ *where* $\lambda_0(r)$ *is the first eigenvalue of*

$$\begin{aligned} -\Delta u &= \lambda r(x) u, \quad x \in \Omega \\ u(x) &= 0, \quad x \in \partial\Omega. \end{aligned} \tag{2.3}$$

Proof. Let $\beta \geq 0$ be a solution of (2.2) where $\lambda \geq \lambda_0(r)$; then

$$-\Delta\beta = \lambda F(x,\beta) \geq \lambda f_0(x) + \lambda r(x)\beta$$

for $x \in \Omega$ with $\beta(x) = 0$ on $\partial\Omega$. Also, $\alpha(x) \equiv 0$ satisfies (2.3) and, since $f_0(x) > 0$ and $\lambda \geq \lambda_0(r) > 0$,

$$-\Delta\alpha(x) = 0 \leq \lambda f_0(x) = \lambda[f_0(x) + r(x)\alpha(x)].$$

Thus, $\beta(x)$ is an upper solution and $\alpha(x)$ is a lower solution of

$$\begin{aligned} -\Delta u &= \lambda[f_0(x) + r(x)u], \quad x \in \Omega \\ u(x) &= 0, \quad x \in \partial\Omega \end{aligned}$$

with $\beta(x) \geq \alpha(x)$. By Theorem 2.1 there exists a solution u where $0 \leq u(x) \leq \beta(x)$. Since $-\Delta u > \lambda r(x)u \geq 0$ and $u(x) \not\equiv 0$ with $u(x) = 0$ on $\partial\Omega$, by the maximum principle, $u(x) > 0$ on Ω.

Let $w(x)$ be a nonnegative eigenfunction corresponding to $\lambda_0(r)$. Integrating $u\Delta w - w\Delta u$ over Ω, we have

$$0 = \int_\Omega (u\Delta w - w\Delta u)\, dx = \int_\Omega \{w[\lambda f_0(x) + \lambda r(x)u(x)] - u[\lambda_0 r(x)w(x)]\}\, dx$$

which implies

$$(\lambda_0 - \lambda)\int_\Omega r(x)u(x)w(x)\, dx = \lambda \int_\Omega w(x) f_0(x) dx > 0,$$

and hence, $\lambda < \lambda_0(r)$ which is a contradiction since we had assumed $\lambda \geq \lambda_0(r)$. □

For example, the Gelfand problem: $-\Delta u = \lambda e^u$ for $x \in \Omega$ and $u(x) = 0$ for $x \in \partial\Omega$, has no solution for $\lambda \geq \lambda_0$ where λ_0 is the first eigenvalue of the problem: $-\Delta u = \lambda u$ for $x \in \Omega$ and $u(x) = 0$ for $x \in \partial\Omega$. This follows since $e^u \geq u + 1$ for $u \geq 0$, and Lemma 2.4 applies.

Note that if $F(x,0) > 0$, $F_u(x,u) > 0$, and $F_{uu}(x,u) \geq 0$ for $(x,u) \in \overline{\Omega} \times [0,\infty)$, then $F(x,u) \geq F(x,0) + F_u(x,0)u$ and Lemma 2.4 applies. If $\lambda \geq \lambda_0(F_u(\cdot,0))$, then $\lambda \notin \Sigma$ of BVP (2.2).

The next lemma is due to Bandle [BAN2] and uses symmetrization and isoperimetric inequalities.

Lemma 2.5 *The solution $w(x)$ of*

$$-\Delta w = 1, \quad x \in \Omega$$
$$w(x) = 0, \quad x \in \partial\Omega$$

is bounded above by $(\frac{V_n}{S_n})^{2/n}(2n)^{-1}$ where V_n and S_n are the n-dimensional volumes of Ω and the unit ball, respectively.

As a consequence of this lemma, we can prove

Theorem 2.6 *Assume there exists a nonnegative nondecreasing function $f_0 \in C^\alpha$ such that $F(x,u) \leq f_0(u)$ for $(x,u) \in \overline{\Omega} \times [0,\infty)$. Suppose that the function $\frac{m}{f(m)}$, $m \geq 0$, assumes its maximum at m_0. If*

$$\lambda_1 = 2nm_0(S_n/V_n)^{2/n}[f_0(m_0)V_n]^{-1},$$

then $[0,\lambda_1] \subseteq \Sigma$ for BVP (2.2).

Proof. Clearly $\alpha(x) \equiv 0$ is a lower solution for (2.2). Select $\lambda \in [0,\lambda_1]$ and consider $\beta(x) = \lambda f_0(m_0)w(x)$. The function $\beta(x)$ is a solution of the boundary value problem

$$-\Delta\beta = \lambda f_0(m_0), \quad x \in \Omega$$
$$\beta(x) = 0, \quad x \in \partial\Omega.$$

In addition, $\beta(x) \geq 0$ on $\overline{\Omega}$ and

$$\beta(x) = \lambda f_0(m_0)w(x) \leq \lambda_1 f_0(m_0)(\tfrac{V_n}{S_n})^{2/n}(2n)^{-1} \leq m_0$$

for $x \in \overline{\Omega}$, where we have used our hypothesis on λ_1. Since

$$\lambda F(x,\beta(x)) \leq \lambda f_0(\beta(x)) \leq \lambda f_0(m_0) = -\Delta\beta,$$

$\beta(x)$ is an upper solution for (2.2); By Theorem 2.1, (2.2) has a solution $u(x) \geq 0$. Consequently, $\lambda \in \Sigma$ and $[0,\lambda_1] \subset \Sigma$. □

We can apply Theorem 2.6 to the Gelfand problem for $\Omega = B_1 \subset \mathbb{R}^n$, $n = 1,2,3$. Since $\frac{m}{f(m)} = me^{-m}$ has a maximum of e^{-1} at $m = 1$, $\lambda_1 = 2n/e$. Let λ_0 be the first eigenvalue for: $-\Delta u = \lambda u$ for $x \in B_1$ and $u = 0$ for $x \in \partial B_1$; then

1. For $n = 1$, $[0, 2/e] \subset \Sigma$ and no solution exists for $\lambda \geq \lambda_0 = \frac{\pi^2}{4}$.

2. For $n = 2$, $[0, 4/e] \subset \Sigma$ and no solution exists for $\lambda \geq \lambda_0 \doteq 5.784$.

3. For $n = 3$, $[0, 6/e] \subset \Sigma$ and no solution exists for $\lambda \geq \lambda_0 \doteq 9.872$.

The next result is due to Kazdan and Warner [KAZ].

Theorem 2.7 *If $F(x,u) > 0$ for $(x,u) \in \Omega \times [0,\infty)$, then there is a $\lambda_0 \in (0,\infty]$ such that a positive solution of BVP (2.2) exists for $\lambda \in (0,\lambda_0)$. No solution exists for $\lambda \geq \lambda_0$ or $\lambda \leq 0$. In addition:*

1. *If $\liminf_{s\to\infty} \frac{F(x,s)}{s} > 0$ uniformly in x, then $\lambda_0 < \infty$.*
2. *If $\lim_{s\to\infty} \frac{F(x,s)}{s} = 0$ uniformly in x, then $\lambda_0 = \infty$.*

Proof. Assume u is a positive solution of BVP (2.2) for $\lambda \leq 0$; then $-\Delta u = \lambda F(x,u) \leq 0$ for $x \in \Omega$ and $u(x) = 0$ for $x \in \partial\Omega$, which imply by the maximum principle that $u \leq 0$ on Ω. This contradiction leads us to conclude that $\lambda > 0$ is necessary for existence of positive solutions to (2.2).

By Lemma 2.3, if there is a positive solution of (2.2) for some $\lambda_1 > 0$, then there is a solution for all $\lambda \in (0,\lambda_1)$. Define

$$P := \{\lambda : \text{BVP (2.2) has a positive solution}\}.$$

To show that $P \neq \emptyset$, we show that (2.2) has a positive solution for some $\lambda_1 > 0$.

The function $\underline{u} \equiv 0$ is a lower solution to BVP (2.2). Let $\overline{u}$ be the solution to

$$-\Delta\overline{u} = 1, \quad x \in \Omega,$$

$$\overline{u}(x) = 0, \quad x \in \partial\Omega.$$

By the maximum principle, $\overline{u}(x) > 0$ on Ω. For $\lambda_1 > 0$ sufficiently small, $-\Delta\overline{u}(x) \geq \lambda_1 F(x,\overline{u}(x))$ for all $x \in \overline{\Omega}$ (where the continuity of F implies $F(x,\overline{u}(x))$ is bounded on $\overline{\Omega}$), and so $\overline{u}$ is an upper solution of (2.2) for $\lambda = \lambda_1$. By Theorem 2.1, there is a solution $u \geq 0$ to BVP (2.2). By the maximum principle, $u(x) > 0$ on $\overline{\Omega}$, so $\lambda_1 \in P$. This proves the first part of the theorem. Define $\lambda_0 = \sup P$.

Proof of part (1). If $\liminf_{s\to\infty} \frac{F(x,s)}{s} > 0$, then we claim that $\lambda_0 < \infty$. The inequality for the lim inf implies the existence of $\alpha \geq 0$ and $\beta > 0$ such that $F(x,s) > \alpha + \beta s$. Thus, if u is a positive solution of (2.2) and if $\psi > 0$ is the eigenfunction associated with the first eigenvalue μ of

$$-\Delta\psi = \mu\psi, \quad x \in \Omega,$$

$$\psi(x) = 0, \quad x \in \partial\Omega,$$

normalized so that

$$\|\psi\|_{L^2(\Omega)} := \sqrt{\int_\Omega \psi^2(x)\,dx} = 1,$$

then

$$\begin{aligned} 0 &= \langle \psi, -\Delta u - \mu u \rangle_{L^2(\Omega)} \\ &= \langle \psi, \lambda F(x,u) - \mu u \rangle_{L^2(\Omega)} \\ &\geq \langle \psi, \lambda \alpha + (\lambda\beta - \mu) u \rangle_{L^2(\Omega)} \end{aligned}$$

which is impossible if $\lambda\beta \geq \mu$. Thus, $\lambda < \frac{\mu}{\beta}$ and $\lambda_0 \leq \frac{\mu}{\beta} < \infty$.

Proof of part (2). If $\lim_{s\to\infty} \frac{F(x,s)}{s} = 0$, then, since $F(x,s) < s$ for s sufficiently large and uniformly in x, one can construct an upper solution $\overline{u}$ for any $\lambda > 0$. The function $\underline{u} \equiv 0$ is always a lower solution. By Theorem 2.1, there is a solution $u \geq 0$, so $\lambda_0 = \sup P = \infty$. □

This theorem gives us additional information for the Gelfand problem and for the perturbed Gelfand problem.

Corollary 2.8 *Given any bounded domain $\Omega \subset \mathbb{R}^n$, there exists $\delta_{FK} \in (0,\infty)$ such that*

1. *for $0 < \delta < \delta_{FK}$, BVP (1.30)-(1.31) has at least one positive solution, and*
2. *for $\delta > \delta_{FK}$, no solution exists.*

Moreover, if Ω is the unit ball in $\mathbb{R}^n$, then

$$\frac{2n}{e} \leq \delta_{FK} \leq \frac{\mu}{e}.$$

where μ is the first eigenvalue of: $-\Delta\psi = \mu\psi$ for $x \in \Omega$ and $\psi(x) = 0$ for $x \in \partial\Omega$.

Proof. The existence of δ_{FK} follows from Theorem 2.7. The lower bound on δ_{FK} follows from Theorem 2.6. Since $e^u \geq eu$ for all $u \geq 0$, the value β in Theorem 2.7 can be chosen to be the number e; the upper bound on δ_{FK} follows. □

The value δ_{FK} is the critical value for the Frank-Kamenetski parameter δ. Frank-Kamenetski [FRA] used this critical value to differentiate between explosive and nonexplosive thermal events. For $\delta > \delta_{FK}$, the nonexistence of a solution for (1.30)-(1.31) was interpreted to mean that an explosion would occur.

Corollary 2.9 *Given any bounded domain $\Omega \subset \mathbb{R}^n$, BVP (1.32)-(1.33) has a solution for any $\delta > 0$.*

Proof. Since for any $\varepsilon > 0$,

$$\frac{1}{u}\exp\left(\frac{u}{1+\varepsilon u}\right) \to 0 \quad \text{as} \quad u \to \infty,$$

the result follows. □

For rather general domains Ω which are open and bounded in $\mathbb{R}^n$ and whose boundaries are of class $C^{2+\alpha}$, there are many existential results for a wide variety of nonlinearities [LIN]. For example, Schuchman [SCU] proved

Theorem 2.10 *Consider boundary value problem (2.2) where $F(x,u)$ is continuously differentiable in $u \in \mathbb{R}^+$ and where $F(x,0) > 0$ for all $x \in \overline{\Omega}$. If there are constants $\alpha > 0$ and $K > 0$ such that*

$$F_u(x,u) \leq K(1+u)^{-(1+\alpha)} \text{ for all } (x,u) \in \overline{\Omega} \times \mathbb{R}^+,$$

then there exists $\lambda_0 > 0$ such that boundary value problem (2.2) has a unique solution for every $\lambda > \lambda_0$.

Some partial multiplicity results are also known [DEF].

Definition 2.3 *A solution $u_{\min}(x)$ of (2.2) is said to be a* minimal solution *of (2.2) if given any other solution $u(x)$ of (2.2), $u_{\min}(x) \leq u(x)$ for all $x \in \Omega$. Similarly one can define a* maximal solution *$u_{\max}(x)$ of (2.2).*

We will use these definitions in later sections in this chapter.

2.2 Radial Symmetry

Symmetrization techniques can be used to simplify a partial differential equation defined on a domain Ω possessing certain symmetry properties. If Ω is a ball in $\mathbb{R}^n$ centered at 0, then one could seek radially symmetric solutions. Although this approach may not produce all solutions to a given problem associated with a given partial differential equation, Gidas, Ni, and Nirenberg [GID1] proved that for a large class of problems, positive solutions are necessarily radially symmetric.

More precisely, for $\Omega = \{x \in \mathbb{R}^n : |x| < R\} = B_R$, let $u \in C^2(\overline{\Omega}, \mathbb{R})$ be a positive solution of

$$\begin{aligned} -\Delta u &= f(u), \quad x \in \Omega \\ u &= 0, \quad x \in \partial\Omega \end{aligned} \tag{2.4}$$

where $f \in C^1(\mathbb{R}, \mathbb{R})$; then u is radially symmetric and radially decreasing. That is, if $r := |x|$, then $u = u(r)$ and $u'(r) < 0$ for $r \in (0, R)$.

This implies that any positive solution of (2.4) is a solution of

$$\begin{aligned} &u'' + \tfrac{n-1}{r}u' + f(u) = 0, \quad 0 < r < R \\ &u'(0) = 0, \quad u(R) = 0. \end{aligned} \tag{2.5}$$

Thus, we need only determine the existence of positive solutions of (2.5).

The assumption that $u > 0$ is necessary. For example, $u(x) = \sin(\pi x)$ is a solution to: $-u'' = \pi^2 u$ for $x \in (-1, 1)$ and $u(\pm 1) = 0$. We have $u(x) > 0$ for $x \in (0, 1)$, but $u(x) < 0$ for $x \in (-1, 0)$. The solution $u(x)$ is not radially symmetric. Even if $u(x) \geq 0$ the full result may not be true. For example, $u(x) = 1 - \cos(2\pi x)$ is a solution to: $-u'' = 4\pi^2(u - 1)$ for $x \in (-1, 1)$ and $u(\pm 1) = 0$. We have $u(x) \geq 0$ (where $u(0) = 0$) and $u(-x) = u(x)$, but $u(x)$ is not radially decreasing. Note that the condition $f(u) \geq 0$ for all u implies that any nontrivial solution is positive (by the maximum principle).

Although the result is stated for $f \in C^1$, this hypothesis can be weakened. The result also holds for any function $f = f_1 + f_2$ where $f_1 \in C^1$ and f_2 is monotone increasing. In particular, the result holds if f is locally Lipschitz continuous.

The proof utilizes maximum principles and the method of moving parallel planes. We first prove a maximum principle which is more delicate to prove than the standard ones. It is a generalization of the Hopf lemma.

Lemma 2.11 *Let Ω^* be a bounded domain whose boundary $\partial\Omega^*$ is of class C^2. Let T be a hyperplane containing the normal to $\partial\Omega^*$ at some point q. Let Ω be that portion of Ω^* which lies on one side of T.*

Let $w \in C^2(\overline{\Omega}, [0, \infty))$ satisfy $\Delta w \leq 0$ for $x \in \Omega$ with $w(q) = 0$. If s is any direction vector at q entering Ω nontangentially, then

$$\frac{\partial w}{\partial s}(q) > 0 \quad or \quad \frac{\partial^2 w}{\partial s^2}(q) > 0$$

unless $w \equiv 0$.

Proof. Without loss of generality, we can orient Ω^* so that the plane T has normal vector $\gamma = (1, 0, \ldots, 0)$. Let Ω be on the side of T which γ points to. Let K_1 be a ball internally tangent to Ω^* at q with radius r_1. Again without loss of generality, translate Ω^* so that the origin $0 \in \mathbb{R}^n$ becomes the center of K_1. Let K_2 be the ball of radius $\frac{1}{2}r_1$ centered at q. Define $K = K_1 \cap K_2 \cap \Omega$. Figure 2.1 illustrates these sets.

Define $z(x) = x_1 \left(e^{-\alpha r^2} - e^{-\alpha r_1^2}\right)$ for $\alpha > 0$; then

$$\begin{aligned}
&\Delta z = 2\alpha x_1 e^{-\alpha r^2}[2\alpha r^2 - (n+2)], \; x \in K, \\
&z(x) > 0, \; x \in K \text{ and} \\
&z(x) = 0, \; x \in T \cup \partial K_1.
\end{aligned}$$

Choose $\alpha = (n+2)/r_1^2$. Since $r \geq \frac{1}{2}r_1$, we have $\Delta z > 0$ on K.

If $w \not\equiv 0$ on Ω, then $w > 0$ on Ω by the maximum principle. By the Hopf Lemma, $\frac{\partial w}{\partial x_1} > 0$ for any point in $\partial K \cap \partial K_2$. Thus, $w > \varepsilon x_1$ on $\partial K \cap \partial K_2$ for some $\varepsilon > 0$. Clearly $w \geq 0$ on $(\partial K \cap \partial K_1) \cup (\partial K \cap T)$. On the other hand, $z \leq x_1$ on $\partial K \cap \partial K_2$.

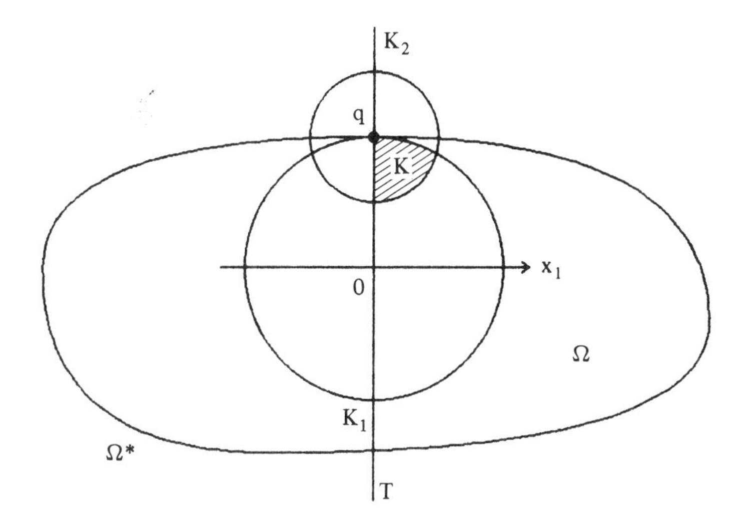

Figure 2.1.

Define $\phi = w - \varepsilon z$; then ϕ satisfies: $\phi(x) \geq 0$ for $x \in \partial K$, $\phi(q) = 0$, and $\Delta\phi < 0$ for $x \in K$. By the maximum principle, $\phi > 0$ on K. At q we have either $\phi_s > 0$ or $\phi_{ss} \geq 0$. But $z_s(q) = 0$ and $z_{ss}(q) > 0$, so either $w_s(q) > 0$ or $w_{ss}(q) > 0$. □

The Method of Moving Parallel Planes. Let $\Omega \subset \mathbb{R}^n$ be a bounded domain with smooth boundary $\partial\Omega$. Let $\lambda \in \mathbb{R}$ and let $\gamma \in \mathbb{R}^n$ be a unit vector. Define $T_\lambda = \{x \in \mathbb{R}^n : \gamma \bullet x = \lambda\}$ to be the hyperplane with normal γ and whose distance from the origin 0 is $|\lambda|$. There is a λ_0 sufficiently large such that $T_{\lambda_0} \cap \overline{\Omega} \neq \emptyset$ and $T_\lambda \cap \overline{\Omega} = \emptyset$ for $\lambda > \lambda_0$. For any $x \in \mathbb{R}^n$, let x^λ be its reflection through T_λ.

Define $\Sigma(\lambda) = \Omega \cap \{x : \gamma \bullet x > \lambda\}$; then $\Sigma(\lambda) = \emptyset$ for $\lambda \geq \lambda_0$ and $\Sigma(\lambda) \neq \emptyset$ for $\lambda < \lambda_0$. The set $\Sigma(\lambda)$ is called an *open cap*. Define $\Sigma'(\lambda)$ to be the reflection of $\Sigma(\lambda)$ through the plane T_λ. An example of these sets is illustrated in Figure 2.2.

For $\lambda < \lambda_0$ with $|\lambda - \lambda_0|$ sufficiently small, we see that $\Sigma'(\lambda) \subseteq \Omega$. Decreasing λ further, we have $\Sigma' \subseteq \Omega$ until either

1. $\Sigma'(\lambda)$ becomes internally tangent to $\partial\Omega$ at some $p \notin T_\lambda$, or

2. T_λ is orthogonal to $\partial\Omega$ at some $q \in T_\lambda \cap \partial\Omega$.

Examples of these conditions are illustrated in Figure 2.3.

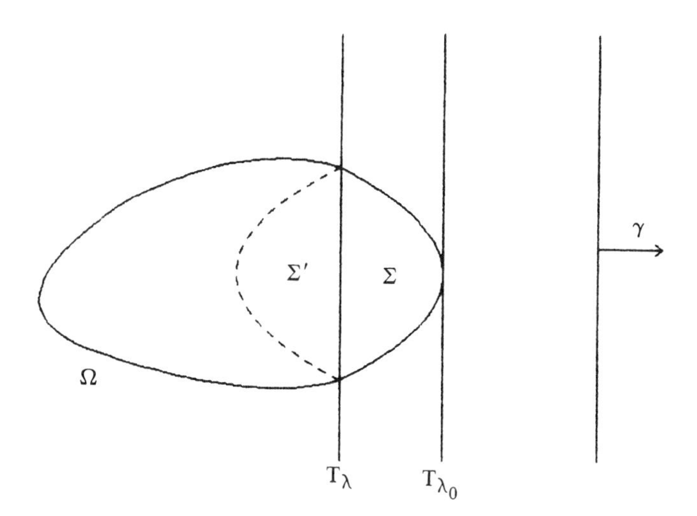

Figure 2.2.

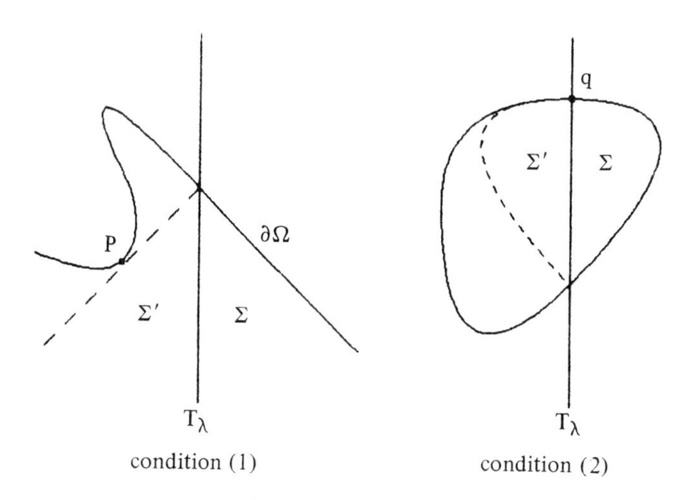

Figure 2.3.

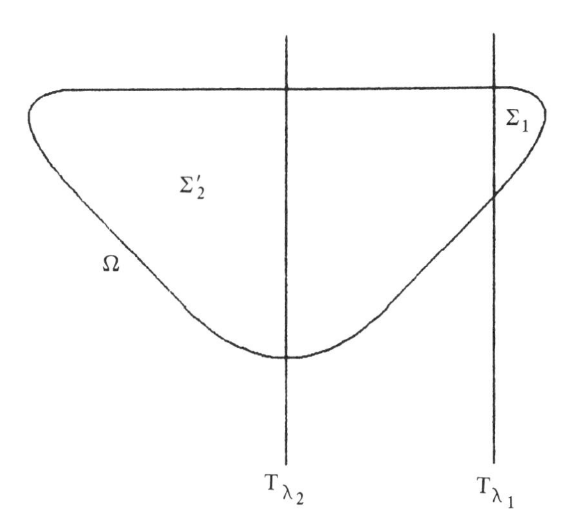

Figure 2.4.

Define

$$\lambda_1 = \sup\{\lambda < \lambda_0 : \text{condition (1) or (2) occurs}\}.$$

The cap $\Sigma(\lambda_1)$ is called the *maximal cap* associated with γ. Note that $\Sigma'(\lambda) \subseteq \Omega$.

If we decrease λ below λ_1, it may be that $\Sigma'(\lambda) \subseteq \Omega$. Define

$$\lambda_2 = \inf\{\lambda < \lambda_0 : \Sigma'(\overline{\lambda}) \subseteq \Omega \ \text{ for } \ \overline{\lambda} \in (\lambda, \lambda_0)\}.$$

The cap $\Sigma(\lambda_2)$ is called the *optimal cap* associated with γ. Note that at λ_2 either (1) or (2) occurs and $\Sigma'(\lambda_2) \subseteq \Omega$. Figure 2.4 illustrates maximal and optimal caps.

For the ensuing arguments we can assume without loss of generality that $\gamma = (1, 0, \ldots, 0) \in \mathbb{R}^n$ and $\lambda_0 = \max\{x_1 : x \in \overline{\Omega}\}$ where $x = (x_1, \ldots, x_n)$. Let λ_1 and λ_2 be defined as above. Define Σ_1 to be the maximal cap associated with γ and denote its reflection through T_{λ_1} by Σ_1'. Define Σ_2 to be the optimal cap associated with γ and denote its reflection through T_{λ_2} by Σ_2'.

For $x_0 \in \partial\Omega$ and $\varepsilon > 0$, define a *neighborhood of x_0 in Ω* by $\Omega_\varepsilon = \Omega \cap B_\varepsilon(x_0)$ where $B_\varepsilon(x_0)$ is the ball of radius ε centered at x_0. Define $S_\varepsilon = \partial\Omega \cap B_\varepsilon(x_0)$. These sets are illustrated in Figure 2.5.

Let $\nu(x) = (\nu_1(x), \ldots, \nu_n(x))$ be the unit outward normal to $\partial\Omega$ at x.

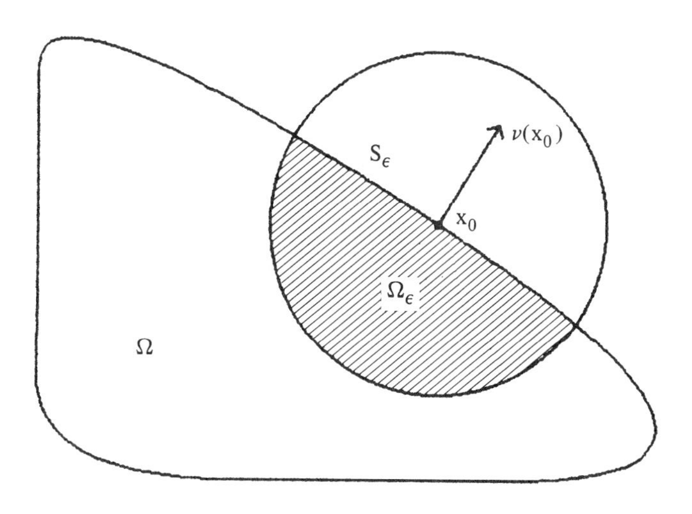

Figure 2.5.

Lemma 2.12 *Let $x_0 \in \partial\Omega$ be such that $\nu(x_0) \bullet \gamma > 0$. Choose $\varepsilon > 0$ sufficiently small so that $\nu(x) \bullet \gamma > 0$ for all $x \in S_\varepsilon$. If $u \in C^2(\overline{\Omega}_\varepsilon)$ satisfies $u_{x_1}(x_0) = 0$, $u(x) = 0$ for $x \in S_\varepsilon$, and $u(x) > 0$ for $x \in \Omega_\varepsilon$, then*

$$\nabla u(x_0) = 0 \quad \textit{and} \quad D^2 u(x_0) = [\Delta u(x_0)]\, \nu(x_0)\nu^t(x_0)$$

where $D^2 u = [u_{x_i x_j}]$ is the $n \times n$ matrix of second derivatives of u.

Proof. On S_ε we know that $u(x) \equiv 0$ and so $\nabla u(x)$ is normal to S_ε at each x. Since S_ε is a smooth $(n-1)$-dimensional manifold, the tangent space $T(x)$ to $x \in S_\varepsilon$ is $(n-1)$-dimensional, say

$$T(x) = \text{span}\,\langle w^1(x), \ldots, w^{n-1}(x)\rangle,$$

where the $w^k(x)$ form an orthonormal set for each x. Consequently, $\nabla u(x) \bullet w^k(x) = 0$ for $k = 1, \ldots, n-1$. Since $\nu(x) \bullet \gamma > 0$ on S_ε it must be that $\gamma \notin T(x)$ and so $\{w^1(x), \ldots, w^{n-1}(x), \gamma\}$ is a basis for $\mathbb{R}^n$ for each x. The basis coefficients for $\nabla u(x_0)$ are given by $\nabla u(x_0) \bullet w^k(x_0) = 0$, $k = 1, \ldots, n-1$, and $\nabla u(x_0) \bullet \gamma = u_{x_1}(x_0) = 0$. Thus, $\nabla u(x_0) = 0$.

Let $x(s)$ be any smooth curve on S_ε such that $x(0) = x_0$. Since $u \equiv 0$ on S_ε, we have

$$y(s)^T \nabla u(x(s)) \equiv 0$$

for any smooth function $y(s) \in T(x(s))$. Differentiating with respect to s gives us

$$y(s)^T D^2 u(x(s)) x'(s) + y'(s)^T \nabla u(x(s)) \equiv 0.$$

One can choose $n-1$ curves $x(s)$ so that at $s = 0$,

$$w^i(x_0)^T D^2 u(x_0) w^j(x_0) = 0, \quad i, j = 1, \ldots, n-1. \tag{2.6}$$

The hypotheses on u guarantee that $\frac{\partial u}{\partial \nu(x)} \leq 0$ for $x \in S_\varepsilon$. Moreover, since $\nabla u(x)$ and $\nu(x)$ are parallel, we have

$$\nu(x(s))^T \nabla u(x(s)) = -|\nabla u(x(s))| =: -p(s)$$

for any smooth curve $x(s)$ on S_ε with $x(0) = x_0$. The function $p(s)$ is nonnegative and differentiable. Thus, at a point where $p = 0$, we must have $p' = 0$. In particular, $p'(0) = 0$ since we had proved $\nabla u(x_0) = 0$. Differentiating with respect to s gives us

$$\left[\frac{d}{ds}\nu(x(s))\right]^T \nabla u(x(s)) + \nu(x(s))^T D^2 u(x(s)) x'(s) = -p'(s).$$

At $s = 0$ we have $\nu(x_0)^T D^2(u(x_0))x'(0) = 0$. The curves $x(s)$ can be chosen to obtain

$$\nu(x_0)^T D^2(u(x_0)) w^j(x_0) = 0, \quad j = 1, \ldots, n-1. \tag{2.7}$$

The set $\{w^1(x_0), \ldots, w^{n-1}(x_0), \nu(x_0)\}$ is orthonormal, so the block matrix

$$Q(x_0) = [w^1(x_0) \mid \cdots \mid w^{n-1}(x_0) \mid \nu(x_0)]$$

is orthogonal, $Q(x_0)e^k = w^k(x_0)$ for $k = 1, \ldots, n-1$, and $Q(x_0)e^n = \nu(x_0)$, where the e^k are the standard Euclidean basis vectors in $\mathbb{R}^n$.

Combining equations (2.6) and (2.7), we obtain

$$m_{ij} = (e^i)^T Q(x_0)^T D^2 u(x_0) Q(x_0) e^j = 0$$

for $i = 1, \ldots, n$ and $j = 1, \ldots, n-1$. Since $D^2 u(x_0)$ is symmetric, from elementary linear algebra we see that

$$Q^T(x_0) D^2 u(x_0) Q(x_0) = \operatorname{diag}\{0, \ldots, 0, m_{nn}(x_0)\}.$$

Similar matrices have the same trace, so

$$\operatorname{trace}(Q(x_0)^T D^2 u(x_0) Q(x_0)) = \operatorname{trace}(D^2 u(x_0)).$$

That is, $m_{nn}(x_0) = \operatorname{trace}(D^2 u(x_0)) = \Delta u(x_0)$. We now have

$$D^2 u(x_0) = Q(x_0) \operatorname{diag}\{0, \ldots, 0, \Delta u(x_0)\} Q^T(x_0) = \nu(x_0) \Delta u(x_0) \nu^T(x_0)$$

which completes the proof. □

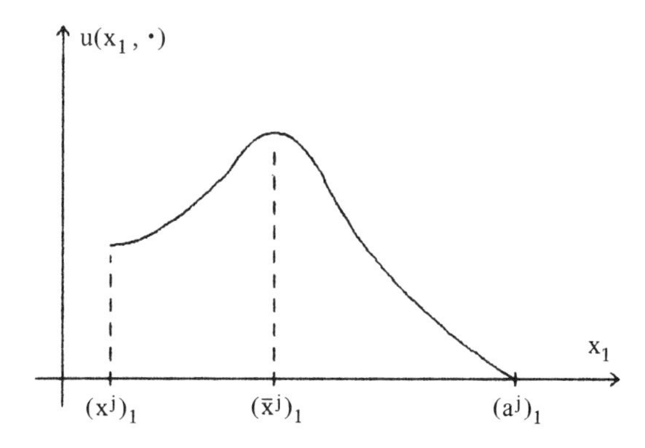

Figure 2.6.

Lemma 2.13 *Let $x_0 \in \partial\Omega$ be such that $\nu(x_0) \bullet \gamma > 0$. Choose $\varepsilon > 0$ sufficiently small so that $\nu(x) \bullet \gamma > 0$ for all $x \in S_\varepsilon$. Assume that $u \in C^2(\overline{\Omega}_\varepsilon)$ satisfies*

1. *$u(x) > 0$ for $x \in \Omega$,*
2. *$\Delta u + f(u) = 0$ for $x \in \Omega_\varepsilon$, and*
3. *$u(x) = 0$ for $x \in S_\varepsilon$;*

then there is a $\delta \in (0, \varepsilon)$ such that $u_{x_1} < 0$ on Ω_δ.

Proof. Since $u > 0$ on Ω and $u = 0$ on S_ε, it is necessary that $\nabla u \bullet w \leq 0$ on S_ε for any vector w such that $\nu \bullet w > 0$. In particular, since $\nu(x) \bullet \gamma > 0$, we must have $u_{x_1}(x) = \nabla u(x) \bullet \gamma \leq 0$ on S_ε.

If the conclusion is false, then there is a sequence $\{x^j\}_{j=1}^{\infty} \subseteq \Omega_\varepsilon$ such that $x^j \to x_0$ as $j \to \infty$ and $u_{x_1}(x^j) \geq 0$. For j large, the interval $I_j \subset \mathbb{R}^n$ in the x_1-direction from x^j to $\partial\Omega$ intersects S_ε at a^j with $u_{x_1}(a^j) \leq 0$. Thus, there exists a sequence $\{\overline{x}^j\}_{j=1}^{\infty} \subseteq \Omega_\varepsilon$ such that $\overline{x}^j \to x_0$ as $j \to \infty$ and $u_{x_1}(\overline{x}^j) = 0$. Figure 2.6 illustrates these values.

By continuity we have

$$u_{x_1}(x_0) = \lim_{j \to \infty} u_{x_1}(\overline{x}^j) = 0. \tag{2.8}$$

By the Mean Value Theorem, there is a sequence $\{\hat{x}^j\}_{j=1}^{\infty} \subseteq I_j$ such that

$$u_{x_1 x_1}(\hat{x}^j) = \frac{u_{x_1}(\overline{x}^j) - u_{x_1}(x_0)}{[\overline{x}^j]_1 - [x_0]_1} = 0,$$

so by continuity we have

$$u_{x_1x_1}(x_0) = \lim_{j\to\infty} u_{x_1x_1}(\hat{x}^j) = 0. \tag{2.9}$$

If $f(0) \geq 0$, then $\Delta u + c(x)u = \Delta u + f(u) - f(0) \leq 0$ on Ω_ε where $c(x)$ is constructed by the Mean Value Theorem. By the Hopf Lemma, $\nabla u(x_0) \bullet w < 0$ for any vector w such that $\nu(x_0) \bullet w > 0$. In particular, $u_{x_1}(x_0) = \nabla u(x_0 \bullet \gamma < 0$, a contradiction to equation (2.8).

If $f(0) < 0$, then by Lemma 2.12 we have $u_{x_ix_j} = [\Delta u(x_0)]\nu_i\nu_j = -f(0)\nu_i\nu_j$ for all i and j. Consequently, $u_{x_1x_1}(x_0) = -f(0)\nu_1^2 \neq 0$, a contradiction to equation (2.9). As a result, our original assumption (that there is no $\delta \in (0,\varepsilon)$ such that $u_{x_1} < 0$ on Ω_δ) is incorrect and the lemma is proved. □

Lemma 2.14 *Suppose there is a $\lambda \in [\lambda_1, \lambda_0)$ such that for $x \in \Sigma(\lambda)$ we have $u_{x_1}(x) \leq 0$ and $u(x) \leq u(x^\lambda)$ with $u(x) \not\equiv u(x^\lambda)$; then $u(x) < u(x^\lambda)$ for $x \in \Sigma(\lambda)$ and $u_{x_1}(x) < 0$ for $x \in \Omega \cap T_\lambda$.*

Proof. For $\gamma = (1, 0, \ldots, 0)$ and $x \in \Sigma(\lambda)$ note that $x^\lambda \in \Sigma'(\lambda)$ is given by $x^\lambda = (2\lambda - x_1, x_2, \ldots, x_n)$.

Define $h(x) := u(x^\lambda)$ for $x \in \Sigma'(\lambda)$ $[x^\lambda \in \Sigma(\lambda)]$; then h satisfies $\Delta h + f(h) = 0$ for $x \in \Sigma'(\lambda)$. Define $w(x) := h(x) - u(x)$ for $x \in \Sigma'(\lambda)$; then $\Delta w + c(x)w = \Delta w + f(h) - f(u) = 0$ for $x \in \Sigma'(\lambda)$ where $c(x)$ is constructed using the Mean Value Theorem. Since $w(x) \leq 0$ for $x \in \Sigma'(\lambda)$ and $w(x) = 0$ for $x \in T_\lambda \cap \Omega$, by the maximum principle we have $w(x) < 0$ for $x \in \Sigma'(\lambda)$, and by Lemma 2.11 we have $\frac{\partial w}{\partial x_1} > 0$ for $x \in T_\lambda \cap \Omega$.

Thus, $u(x^\lambda) = h(x) < u(x)$ for $x \in \Sigma'(\lambda)$ and $0 < w_{x_1} = h_{x_1} - u_{x_1} = -2u_{x_1}$ for $x \in T_\lambda \cap \Omega$. This implies that $u_{x_1} < 0$ for $x \in T_\lambda \cap \Omega$ and $u(x^\lambda) < u(x)$ for $x \in \Sigma'(\lambda)$ [which implies $u(x) < u(x^\lambda)$ for $x \in \Sigma(\lambda)$]. □

Lemma 2.15 *Let $H(\lambda) = \{x \in \mathbb{R}^n : x_1 > \lambda\}$. Let $u(x) > 0$ on Ω, $u \in C^2(\overline{\Omega} \cap H(\lambda_1))$, and $u(x) = 0$ on $\partial\Omega \cap H(\lambda_1)$. For any $\lambda \in (\lambda_1, \lambda_0)$ we have $u_{x_1}(x) < 0$ and $u(x) < u(x^\lambda)$ for $x \in \Sigma(\lambda)$.*

Proof. By Lemma 2.13, for λ close to λ_0 with $\lambda < \lambda_0$ we have

$$u_{x_1}(x) < 0 \text{ and } u(x) < u(x^\lambda), \quad x \in \Sigma(\lambda). \tag{2.10}$$

Decrease λ until a critical value

$$\mu = \inf\{\overline{\lambda} \in [\lambda_1, \lambda_0) : \quad (2.10) \text{ holds for } \lambda \in (\overline{\lambda}, \lambda_0)\}$$

is reached. Equation (2.10) then holds for $\mu < \lambda < \lambda_0$ and, for $\lambda = \mu$ (using continuity),

$$u_{x_1}(x) \leq 0 \text{ and } u(x) \leq u(x^\lambda), \quad x \in \Sigma(\mu). \tag{2.11}$$

We claim that $\mu = \lambda_1$. Assume not; then $\mu > \lambda_1$. For any $x_0 \in \partial\Sigma(\mu) \setminus T_\mu$ we have $x_0^\mu \in \Omega$. Since $0 = u(x_0) < u(x_0^\mu)$ we see that $u(x^\mu) \not\equiv u(x)$ in $\Sigma(\mu)$ and so Lemma 2.14 applies. Thus, $u(x) < u(x^\mu)$ for $x \in \Sigma(\mu)$ and $u_{x_1} < 0$ for $x \in \Omega \cap T_\mu$ and equation (2.10) holds for $\lambda = \mu$.

Since $u_{x_1} < 0$ on $\Omega \cap T_\mu$, by Lemma 2.13 there is an $\varepsilon > 0$ such that $u_{x_1} < 0$ on $\Omega \cap H(\mu - \varepsilon)$. By the definition of μ, there are sequences $\{\Lambda_j\}_{j=1}^\infty$ and $\{x_j\}_{j=1}^\infty$ with $\Lambda_j \in (\mu - \varepsilon, \mu)$ and $x_j \in \Sigma(\Lambda_j)$ satisfying

$$\Lambda_j \uparrow \mu \text{ as } j \to \infty \text{ and } u(x_j) \geq u(x_j^{\Lambda_j}) \tag{2.12}$$

By compactness of $\overline{\Sigma(\lambda_1)}$ there is a subsequence $\{x_{j_k}\}_{k=1}^\infty$ such that $x_{j_k} \to x \in \Sigma(\mu)$. Thus,

$$x_{j_k}^{\Lambda_{j_k}} \to x^\mu \in \Sigma'(\mu) \text{ and } u(x) \geq u(x^\mu). \tag{2.13}$$

But $x \in \partial\Sigma(\mu)$ since (2.10) holds for $\lambda = \mu$. If $x \notin T_\mu$, then $x^\mu \in \Omega$ and by (2.13), $0 = u(x) \geq u(x^\mu)$ which is a contradiction to $u > 0$ on Ω. Thus, $x \in T_\mu$ and $x = x^\mu$.

For k sufficiently large, the line segment joining x_{j_k} and $x_{j_k}^{\Lambda_{j_k}}$ is in Ω. From (2.12) and the Mean Value Theorem, there is a y_{j_k} such that $u_{x_1}(y_{j_k}) \geq 0$. Letting $k \to \infty$ we obtain $u_{x_1}(x) \geq 0$ where $x \in T_\mu$, a contradiction since (2.10) holds for $\lambda = \mu$. Thus, our assumption that $\mu > \lambda_1$ is incorrect. In fact, $\mu = \lambda_1$ and (2.10) is valid for all $\lambda \in (\lambda_1, \lambda_0)$. □

Corollary 2.16 *If $u_{x_1}(x) = 0$ for some $x \in \Omega \cap T_{\lambda_1}$, then u is symmetric in T_{λ_1} and*

$$\Omega = \Sigma(\lambda_1) \cup \Sigma'(\lambda_1) \cup [T_{\lambda_1} \cap \Omega].$$

Proof. If $u_{x_1}(x) = 0$ for some $x \in \Omega \cap T_{\lambda_1}$, then by Lemma 2.14 we have $u(x) \equiv u(x^{\lambda_1})$ for $x \in \Sigma(\lambda_1)$. This implies that u is symmetric relative to T_{λ_1}. Since $u(x) > 0$ in $\Sigma(\lambda_1)$ and $u = 0$ on $\partial\Omega$, we conclude that $\Omega = \Sigma(\lambda_1) \cup \Sigma'(\lambda_1) \cup [T_{\lambda_1} \cap \Omega]$. □

We now give the proof of the main result on radial symmetry stated in the introduction to this section.

Theorem 2.17 *For $\Omega = \{x \in \mathbb{R}^n : |x| < R\}$, let $u \in C^2(\overline{\Omega})$ be a positive solution of BVP (2.4) where $f \in C^1$; then $u = u(r)$ where $r = |x|$ and $u'(r) < 0$ for $r \in (0, R)$.*

Proof. By Lemma 2.15 and Corollary 2.16, $u_{x_1} < 0$ for all x with $x_1 > 0$. This implies that $u_{x_1}(x) > 0$ for $x_1 < 0$. Consequently, $u_{x_1}(x) = 0$ for $x_1 = 0$. By Corollary 2.16, u is symmetric in x_1. Since the direction vector γ is arbitrary, the argument above works for any direction. It follows that u is radially symmetric and $u_r < 0$ for $0 < r < R$. □

We now turn our attention to a related symmetry result for an overdetermined boundary value problem. The motivation comes from a problem solved by Serrin [SER] (where $f(u) \equiv 1$).

Theorem 2.18 *Let $\Omega \subset \mathbb{R}^n$ be a bounded domain whose boundary $\partial\Omega$ is of class C^2. Let u be a positive solution of*

$$\Delta u + f(u) = 0, \quad x \in \Omega$$

$$u(x) = 0, \ \frac{\partial u}{\partial n} = c, \quad x \in \partial\Omega,$$

where $f \in C^1$, then $\Omega = \{x \in \mathbb{R}^n : |x| < R\}$. Moreover, u is radially symmetric and radially decreasing.

Proof. As in the proof of Theorem 2.17, without loss of generality let $\gamma = (1, 0, \ldots, 0)$. Using the notation of that theorem, we will show that Ω is symmetric about the hyperplane T_λ. In this case, since the direction γ is arbitrary and since Ω is simply connected, it must be that Ω is a ball. Consequently, Theorem 2.17 applies and so the solution u must be radially symmetric and radially decreasing.

Define (as in Lemma 2.14) the function $h(x) = u(x^{\lambda_1})$ for $x \in \Sigma'(\lambda_1)$; then h satisfies

$$\Delta h + f(h) = 0, \quad x \in \Sigma'(\lambda_1),$$

$$h = u, \quad x \in T_{\lambda_1} \cap \partial\Sigma'(\lambda_1), \quad \text{and}$$

$$h = 0, \quad x \in \partial\Sigma'(\lambda_1) \setminus T_{\lambda_1}.$$

Set $w = h - u$ for $x \in \Sigma'(\lambda_1)$. The maximum principle implies that either $w > 0$ or $w \equiv 0$ on Σ'.

If $w \equiv 0$ on Σ', then Ω is symmetric about T_{λ_1} and the proof is complete. If $w > 0$ on Σ', then $h(x) > u(x)$ for all $x \in \Sigma'$. Recall that $\Sigma'(\lambda_1)$ is either

1. internally tangent to $\partial\Omega$ at $p \in T_{\lambda_1}$, or,

2. T_{λ_1} is orthogonal to $\partial\Omega$ at some $q \in T_{\lambda_1} \cap \partial\Omega$.

If (1) holds, then $w(p) = 0$ and $w_\nu(p) < 0$ for an outward unit normal vector ν. But $w_\nu(p) = h_\nu(p) - u_\nu(p) = c - c = 0$, a contradiction. It must be that condition (2) holds. We will show that w has a zero of second order at q. Lemma 2.11 will then provide us with a contradiction so that in fact $w \equiv 0$.

Since $\partial\Omega$ is of class C^2, consider a rectangular coordinate system with origin at q for which $\partial\Omega$ can be represented locally by $x_n = \phi(x_1, \ldots, x_{n-1})$ where ϕ is a C^2 function, and where the x_n-axis is in the direction of the normal vector $\nu(q)$.

Since $u \in C^2(\overline{\Omega})$, $u \equiv 0$ on $\partial\Omega$ can be expressed as

$$u(x_1, \ldots, x_{n-1}, \phi(x_1, \ldots, x_{n-1})) \equiv 0. \tag{2.14}$$

Let $\Phi(x_1, \ldots, x_n) = x_n - \phi(x_1, \ldots, x_{n-1})$; then $\partial\Omega$ is represented by $\Phi = 0$, so $\nabla\Phi$ is normal to $\partial\Omega$. Consequently,

$$\nu(x) = \frac{\nabla\Phi}{|\nabla\Phi|} = \frac{e_n - \nabla\phi}{|e_n - \nabla\phi|} = \frac{e_n - \nabla\phi}{\sqrt{1 + |\nabla\phi|^2}}, \quad x \in \partial\Omega$$

where $e_n = (0, \ldots, 0, 1)$. Moreover, $\nu(q) = e_n$, so $\nabla\phi(q) = 0$. The quantity $c = \frac{\partial u(x)}{\partial\nu(x)} = \nu(x) \bullet \nabla u(x)$ on $\partial\Omega$ can be expressed as

$$u_{x_n} - \sum_{k=1}^{n-1} u_{x_k}\phi_{x_k} = c\left[1 + \sum_{k=1}^{n-1} \phi_{x_k}^2\right]^{1/2} \tag{2.15}$$

where x_n is to be replaced by $\phi(x_1, \ldots, x_{n-1})$. Differentiating (2.14) with respect to x_i yields

$$u_{x_i} + u_{x_n}\phi_{x_i} = 0,$$

for $i = 1, \ldots, n-1$. Evaluating these equations at q produces $u_{x_i}(q) = 0$ for $i = 1, \ldots, n-1$. These conditions and equation (2.15) imply $u_{x_n}(q) = c$. Differentiating (2.14) with respect to x_i followed by x_j yields

$$u_{x_i x_j} + u_{x_n}\phi_{x_j} + u_{x_n}\phi_{x_i x_j} + (u_{x_n x_j} + u_{x_n x_n}\phi_{x_j})\phi_{x_i} = 0,$$

for $i, j = 1, \ldots, n-1$. Evaluating these equations at q produces $u_{x_i x_j}(q) = -c\phi_{x_i x_j}(q)$ for $i, j = 1, \ldots, n-1$. Differentiating (2.15) with respect to x_i yields

$$u_{x_n x_i} = \sum_{k=1}^{n-1} \left[u_{x_k}\phi_{x_k x_i} + (u_{x_k x_i} + u_{x_n}\phi_{x_i})\,\phi_{x_k}\right] + 2c\sum_{k=1}^{n-1} \phi_{x_k}\phi_{x_k x_i}$$

for $i = 1, \ldots, n-1$. Evaluating at q produces $u_{x_n x_i}(q) = 0$ for $1 \le i \le n-1$.

Finally,

$$u_{x_n x_n} = \Delta u - \sum_{i=1}^{n-1} u_{x_i x_i} = -f(u) + c\Delta\phi$$

for all $x \in \partial\Omega$ near q, in particular at q. Therefore, we have determined all first and second derivatives of u at q.

Since $\Sigma' \subseteq \Omega$, we have $\phi_{1\ell}(q) = 0$ for $\ell = 2, \ldots, n-1$. Also,

$$h(x_1, x_2, \ldots, x_n) = u(-x_1, x_2, \ldots, x_n)$$

so first and second derivatives of u and h agree at q. Applying Lemma 2.11 to $w = h - u$ on $\Sigma'(\lambda_1)$ gives us $w_s(q) > 0$ or $w_{ss}(q) > 0$. This contradicts u and h having the same first and second partials at q. □

2.3 Multiplicity in Special Domains

For $\Omega = B_1 \subset \mathbb{R}^n$, we can get very precise results for the Gelfand problem

$$\begin{aligned} -\Delta u &= \delta e^u, \quad x \in B_1 \\ u(x) &= 0, \quad x \in \partial B_1. \end{aligned} \tag{2.16}$$

To find solutions $u(x) \in C^2(\overline{\Omega})$, because of the maximum principle all solutions are positive, and hence by Theorem 2.17, all solutions are radially symmetric. One can equivalently look for positive solutions $u(r) \in C^2[0,1]$ of

$$u'' + \frac{n-1}{r}u' + \delta e^u = 0, \quad 0 < r < 1 \tag{2.17}$$

with boundary conditions

$$u'(0) = 0,\ u(1) = 0 \ \text{ or } \ u(0) = \alpha,\ u(1) = 0 \tag{2.18}$$

with $u'(1) = -\beta < 0$.

The often quoted multiplicity result due to Joseph and Lundgren [JOS] is

Theorem 2.19 *Consider BVP (2.17)-(2.18). The following existence results hold:*

1. $n = 1$*: There exists* $\delta_{FK} > 0$ *such that*
   (a) *for each* $\delta \in (0, \delta_{FK})$*, there are two solutions,*
   (b) *for* $\delta = \delta_{FK}$*, there is a unique solution, and*
   (c) *for* $\delta > \delta_{FK}$*, there are no solutions.*
2. $n = 2$*: Let* $\delta_{FK} = 2$*; then*
   (a) *for each* $\delta \in (0, \delta_{FK})$*, there are two solutions,*
   (b) *for* $\delta = \delta_{FK}$*, there is a unique solution, and*
   (c) *for* $\delta > \delta_{FK}$*, there are no solutions.*
3. $3 \le n \le 9$*: Let* $\tilde{\delta} = 2(n-2)$*; then there exists* $\delta_{FK} > \tilde{\delta}$ *such that*
   (a) *for* $\delta = \delta_{FK}$*, there is a unique solution,*
   (b) *for* $\delta > \delta_{FK}$*, there are no solutions,*
   (c) *for* $\delta = \tilde{\delta}$*, there is a countable infinity of solutions, and*
   (d) *for* $\delta \in (0, \delta_{FK}) \setminus \{\tilde{\delta}\}$*, there is a finite number of solutions.*
4. $n \ge 10$*: Let* $\delta_{FK} = 2(n-2)$*; then*
   (a) *for* $\delta \ge \delta_{FK}$*, there are no solutions, and*
   (b) *for each* $\delta \in (0, \delta_{FK})$*, there is a unique solution.*

Proof. Recall that $\alpha = u(0)$ and $\beta = -u'(1)$. For $n = 1$, (2.17) can be solved by integration to obtain

$$u(r) = \alpha - 2\ln\cosh\left(\frac{1}{2}r\sqrt{2\delta e^{\alpha}}\right)$$

where α, β, and δ are related by

$$\beta = \sqrt{\beta^2 + 2\delta}\,\tanh\left(\frac{1}{2}\sqrt{\beta^2 + 2\delta}\right)$$

and

$$\delta = \frac{1}{2}e^{-\alpha}\left[\ln\left(\frac{1+\sqrt{1-e^{-\alpha}}}{1-\sqrt{1-e^{-\alpha}}}\right)\right]^2.$$

For $n = 2$, (2.17) can also be solved by making the change of variables $r = e^{-t}$ and $w(t) = u(r) - 2t$ to obtain $\ddot{w} + \delta e^{w} = 0$. We obtain

$$u(r) = \alpha - 2\ln\left(1 + \frac{1}{8}\delta e^{\alpha} r^2\right)$$

where the parameters are related by

$$\beta^2 - 4\beta + 2\delta = 0 \text{ and } \delta = 8(e^{\frac{1}{2}\alpha} - e^{-2\alpha}).$$

For $n \geq 3$, let $t_1 = \frac{1}{2}\ln\left[\frac{2(n-2)}{\delta e^{\alpha}}\right]$, $r = e^{-(t-t_1)}$, and $u(r) = \alpha + 2t + z(t)$; then (2.17) becomes

$$\begin{aligned} &\tfrac{1}{n-2}\ddot{z} - \dot{z} + 2e^{z} - 2 = 0, \quad t_1 < t < \infty \\ &z(\infty) = -\infty, \ \dot{z}(\infty) = -2 \end{aligned} \tag{2.19}$$

with compatibility condition $z(t_1) = -\alpha - 2t_1$. We analyze this problem in the phase plane. Let $y(t) = \dot{z} + 2$ and $x(t) = 2(n-2)\exp(z(t))$; then problem (2.19) is equivalent to

$$\begin{aligned} &\dot{x} = x(y-2), \quad \dot{y} = (n-2)y - x, \quad t_1 < t < \infty, \\ &x(\infty) = y(\infty) = 0 \end{aligned} \tag{2.20}$$

with compatibility condition $t_1 = \frac{1}{2}\ln\left[\frac{2(n-2)}{x(t_1)e^{\alpha}}\right]$; also $\delta = x(t_1)$ and $\beta = y(t_1)$.

The two-dimensional system (2.20) has critical points at $(0,0)$ and $(2(n-2), 2)$. We now prove that there exists a unique heteroclinic orbit joining these two critical points.

Existence. Choose $x_0 \in (0, 2(n-2))$. Consider the following sets:

$$L_+ = \{(x,y) : y = \tfrac{x}{n-2},\, 0 < x \le x_0\},$$
$$L_- = \{(x,y) : y = 0,\, 0 < x \le x_0\},$$
$$L = \{(x,y) : x = x_0,\, 0 \le y \le \tfrac{x_0}{n-2}\}, \text{ and}$$
$$T = \{(x,y) : 0 < x < x_0,\, 0 < y < \tfrac{x}{n-2}\}.$$

The triangular domain T is open and contains no critical points of (2.20). Define the subsets of L,

$$E_+ = \{(x_0, y) \in L : \Phi_t(y) \text{ exits } T \text{ through } L_+\}$$

and

$$E_- = \{(x_0, y) \in L : \Phi_t(y) \text{ exits } T \text{ through } L_-\},$$

where $\Phi_t(y)$ indicates the flow of (2.20) in the xy-plane with initial data $\Phi_{t_0}(y) = (x_0, y)$.

The vector field for (2.20) points strictly outward from T on both L_+ and L_-. In particular, $(x_0, x_0/(n-2)) \in E_+$ and $(x_0, 0) \in E_-$, so these sets are nonempty. By continuous dependence, the sets E_+ and E_- are open sets (relative to L). Moreover, $\overline{E}_+ \cap E_- = \emptyset$ and $E_+ \cap \overline{E}_- = \emptyset$. Since L is a connected set, $L \ne E_+ \cup E_-$. That is, there must be at least one point (x_0, y_0) whose flow $\Phi_t(y_0)$ meets $(0,0)$ at $t = \infty$.

Uniqueness. In the region T, any solution to (2.20) has the property $\dot{x}(t) < 0$, so by the Inverse Function Theorem, one can think of $t = t(x)$ and $y = y(x)$. In T, system (2.20) can be written as

$$\frac{dy}{dx} = \frac{(n-2)y - x}{x(y-2)}, \quad y(0) = 0 \tag{2.21}$$

for $0 < x < x_0$. Suppose that (2.21) has two solutions, $y_1(x)$ and $y_2(x)$. Define $D(x) = y_1(x) - y_2(x)$. Equation (2.21) has a unique solution for any initial data where $x > 0$, so $D \ne 0$ in T. Without loss of generality, say $D > 0$ for $x > 0$. It can be shown that

$$\frac{1}{D}\frac{dD}{dx} = \frac{x - 2(n-2)}{x(y_1 - 2)(y_2 - 2)} \tag{2.22}$$

As $x \to 0^+$, $y_1(x) \to 0$ and $y_2(x) \to 0$. Consequently, the right-hand side of (2.22) approaches $-\infty$. For $x > 0$ and sufficiently small, we have

$$\frac{d}{dx}\ln D(x) = \frac{1}{D}\frac{dD}{dx} \le -\varepsilon < 0.$$

An integration yields $D(x) \ge D(x_1)\exp(\varepsilon(x_1 - x))$ for $0 < x \le x_1$. However, this implies that $D(0) > 0$, a contradiction since $D(0) = 0$.

We have proved that there is a unique solution to (2.20) for $t_1 < t < \infty$. We now want to determine the behavior of this solution as $t \to -\infty$. One can show that there is a rectangle $R = (0, \overline{x}) \times (0, \overline{y})$ such that the solution $(x(t), y(t)) \in R$ for all $t \in (-\infty, t_1)$, and so this solution must converge either to a limit cycle or a critical point as $t \to -\infty$. The vector field for the system rules out the critical point $(0, 0)$.

At the critical point $(2(n-2), 2)$, the linearization of (2.20) is

$$\frac{d}{dt}\begin{bmatrix} x - 2(n-2) \\ y - 2 \end{bmatrix} = \begin{bmatrix} 0 & 2(n-2) \\ -1 & n-2 \end{bmatrix}\begin{bmatrix} x - 2(n-2) \\ y - 2 \end{bmatrix}.$$

The eigenvalues for the linearization are

$$\lambda = \frac{1}{2}\left[(n-2) \pm \sqrt{(n-2)(n-10)}\right].$$

If $2 < n < 10$, then the eigenvalues are complex-valued with positive real parts; the critical point is a spiral node. If $n = 10$, then the eigenvalue is unique and positive; the critical point is an unstable node. For $n > 10$, the eigenvalues are distinct positive real numbers, so the critical point is an unstable node (and one eigenspace is dominant). In any case, the same behavior holds locally for (2.20) at the critical point as in the linear case.

One can see from the vector field for (2.20) that either $(x(t), y(t)) \to (2(n-2), 2)$ as $t \to -\infty$ or $(x(t), y(t))$ spirals about $(2(n-2), 2)$. In this last case, the orbit cannot spiral to a non-constant periodic orbit. If there were such a periodic orbit $\partial\Omega$ enclosing a region Ω, then on this orbit the solution $(x(t), y(t))$ would satisfy

$$-\dot{y}dx + \dot{x}dy = 0 \quad \text{or} \quad -\frac{\dot{y}}{x}dx + \frac{\dot{x}}{x}dy = 0.$$

If $F(x,y) = \frac{\dot{x}}{x} = y - 2$ and $G(x,y) = \frac{\dot{y}}{x} = \frac{(n-2)y}{x} - 1$, then by Green's Theorem,

$$0 = \oint_{\partial\Omega} (-G\,dx + F\,dy) = \int_{\Omega} (F_x + G_y)\,dA = \int_{\Omega} \frac{n-2}{x}\,dA > 0,$$

a contradiction. Thus, there are no limit cycles for (2.19). Moreover, there is a unique heteroclinic orbit connecting the two critical points $(0, 0)$ and $(2(n-2), 2)$.

We summarize our observations in terms of the following (δ, β) bifurcation diagrams shown in Figure 2.7. Each point $(x(t_1), y(t_1))$ on the bifurcation curve is equal to a point (δ, β) via the change of variables (2.19) and (2.20), and via the compatability condition $t_1 = \frac{1}{2}\ln\left[\frac{2(n-2)}{\delta e^{\alpha}}\right]$. That is, $z(t_1) = -\alpha - 2t_1$ implies $x(t_1) = 2(n-2)\exp(z(t_1)) = \delta$, and $\dot{z}(t_1) = -2 + \beta$ implies $y(t_1) = \beta$. This pair (δ, β) provides us with a solution

$$u(r) = \alpha + 2t + \ln\left(\frac{x(t)}{2(n-2)}\right), \quad r = e^{-(t-t_1)}$$

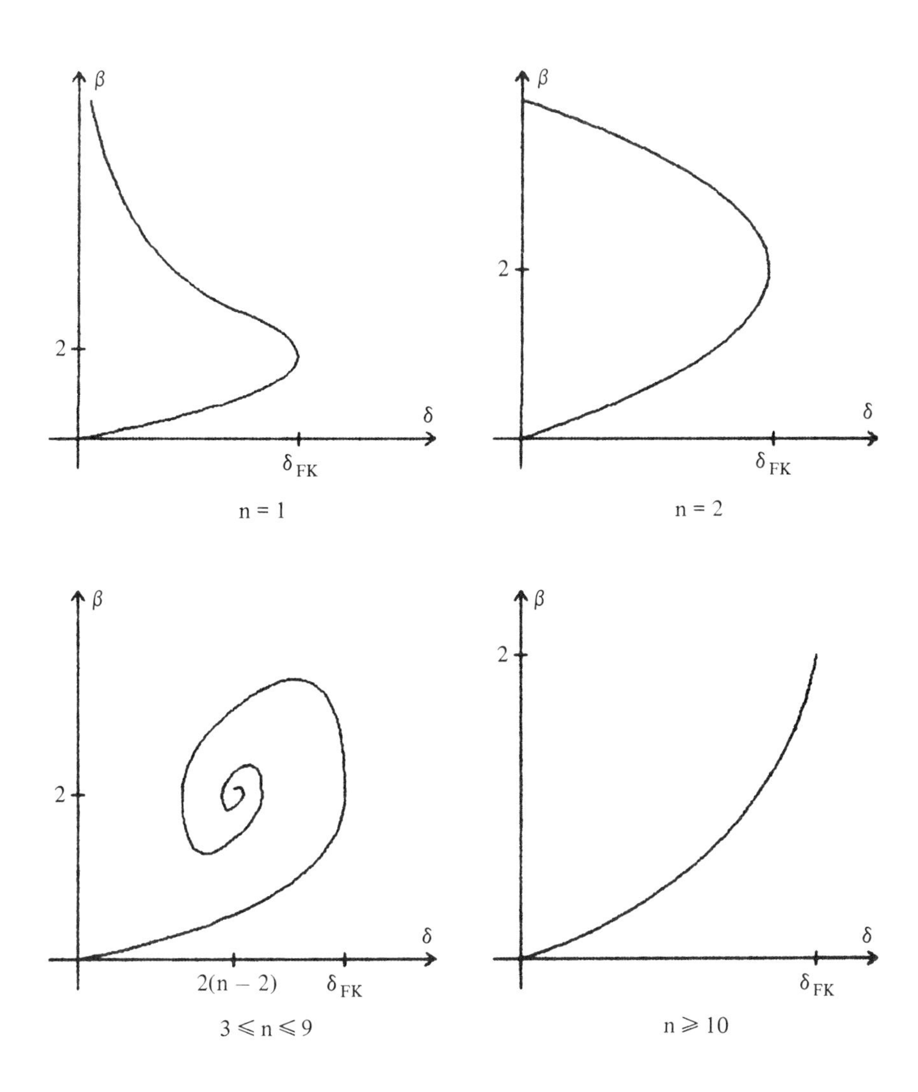
β
2
δ
δ_{FK}
n = 1
β
2
δ
δ_{FK}
n = 2
β
2
δ
2(n − 2)
δ_{FK}
3 ⩽ n ⩽ 9
β
2
δ
δ_{FK}
n ⩾ 10

Figure 2.7.

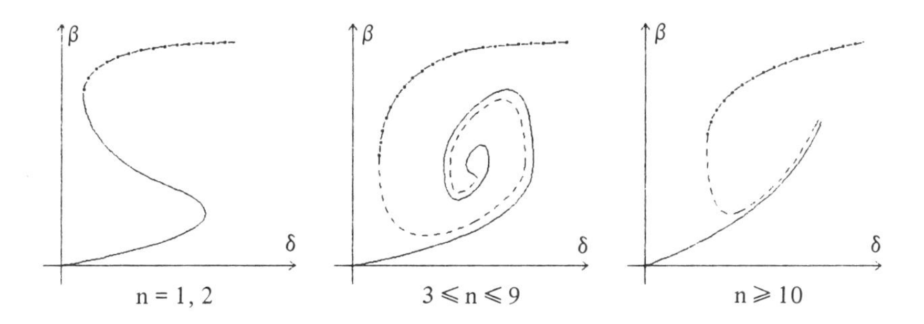

Figure 2.8.

such that $u(0) = \alpha$, $u'(0) = 0$, $u(1) = 0$, and $u'(1) = -\beta$. One can determine the multiplicity of solutions for a given $\overline{\delta}$ by observing how many times the vertical line $\delta = \overline{\delta}$ intersects the heteroclinic orbit. □

For the perturbed Gelfand problem

$$\begin{aligned} -\Delta u &= \delta \exp(\tfrac{u}{1+\varepsilon u}), \quad x \in \Omega \\ u(x) &= 0, \quad x \in \partial\Omega \end{aligned} \tag{2.23}$$

where $\Omega = B_R = \{x \in \mathbb{R}^n : |x| < R\}$, Dancer [DAN1] proved:

Theorem 2.20 *For any $\varepsilon > 0$ and $\delta > 0$, there exists at least one and at most finitely many solutions to BVP (2.23).*

The proof uses perturbation arguments and is quite involved and technical. We only illustrate the results in Figure 2.8. The value ε is positive and chosen sufficiently close to 0.

Using the terminology of Dancer, the solutions represented by the solid line (—) in Figure 2.8 are called small–small solutions. Those solutions represented by the broken line (- -) are called large–small solutions. Those solutions represented by dotted line (– · –) are called large solutions. The shape of the bifurcation curves indicated in Figure 2.8 may not be exactly correct. The branch representing large–small solutions is arbitrarily close to the small–small solution branch for $\varepsilon > 0$ but small. We use this fact in Section 2.4 on solution profiles.

2.4 Solution Profiles

We first consider the Gelfand problem

$$-\Delta u = \delta e^u, \quad x \in B_1$$
$$u = 0, \quad x \in \partial B_1$$

which by Theorem 2.17 is equivalent to

$$\begin{aligned} & u'' + \tfrac{n-1}{r} u' + \delta e^u = 0, \quad 0 < r < 1 \\ & u'(0) = 0, \quad u(1) = 0. \end{aligned} \tag{2.24}$$

We will define a solution u(r) of (2.24) to be *bell-shaped* if it has a unique point of inflection for $r \in (0,1)$. We can then prove the following result.

Theorem 2.21 *Consider BVP (2.24).*

1. *For $n = 1$, all solutions are concave on $[0,1]$.*
2. *For $n = 2$:*
   - *(a) If $\delta \in (0, \delta_{FK})$, then the minimal solution is concave on $[0,1]$ and the maximal solution is bell-shaped.*
   - *(b) If $\delta = \delta_{FK} = 2$, then the solution is concave on $[0,1)$ with $u''(1) = 0$.*
3. *For $n \geq 3$, there exists $\overline{\delta} < \delta_{FK}$ such that:*
   - *(a) If $\delta = \overline{\delta}$, then the minimal solution is concave on $[0,1)$ with $u''(1) = 0$.*
   - *(b) If $\overline{\delta} < \delta \leq \delta_{FK}$, then all solutions are bell-shaped.*
   - *(c) If $0 < \delta < \overline{\delta}$, then the minimal solution is concave on $[0,1]$ and all other solutions are bell-shaped.*

Proof. For $n = 1$ we see that $u''(r) < 0$ on $[0,1]$ and the concavity is obvious.

For $n \geq 2$, note that $u''(0) = -\frac{\delta}{n} e^{\alpha} < 0$ and that $u''(1) = (n-1)\beta - \delta$, so $\operatorname{sgn}[u''(1)] = \operatorname{sgn}[(n-1)\beta - \delta]$.

If the points of inflection are unique (if they exist) and if the bifurcation curve intersects $\beta = L(\delta) = \frac{\delta}{n-1}$ uniquely on the minimal branch, then our assertions (2) and (3) hold. For if $(\delta, \beta) \in \mathcal{D}$, where $\mathcal{D}$ is the bifurcation curve, satisfies $\beta > L(\delta)$, then $u''(1) > 0$ and $u''(0) < 0$ imply there exists $R \in (0,1)$ such that $u''(R) = 0$ and $(R, u(R))$ is a point of inflection. By the uniqueness of inflection points, the solution u(r) corresponding to (δ, β) is bell-shaped. If $(\delta, \beta) \in \mathcal{D}$ satisfies $\beta < L(\delta)$, then $u''(1) < 0$ and $u''(0) < 0$ imply no inflection points or more than one. Uniqueness would rule out the latter case.

To complete the proof, we must prove that

(i) $\mathcal{D}$ intersects $\beta = L(\delta)$ uniquely, and

(ii) there is at most one inflection point for the solution u.

We will prove (i) first. For $n = 2$, (i) is immediate since $\mathcal{D} = \{(\delta, \beta) : \beta^2 - 4\beta + 2\delta = 0\}$. For $n \geq 3$, (i) is a consequence of a sequence of lemmas.

Lemma 2.22 *Let $u(x)$ be a solution of (2.24); then*

$$\int_{B_1} [(2-n)u + 2n]\delta e^u dx = (\beta^2 + 2\delta) w_n$$

where w_n is the surface area of the unit sphere $B_1 \subset \mathbb{R}^n$.

Proof. Note that

$$\Delta\left(r\frac{\partial u}{\partial r}\right) = r\frac{\partial(\Delta u)}{\partial r} + 2\Delta u = -\delta e^u\left(r\frac{\partial u}{\partial r} + 2\right). \qquad (2.25)$$

Define $v = \delta(u-2)e^u$; then $v = -2\delta$ for $x \in \partial B_1$ and $r\frac{\partial v}{\partial r} = \delta(u-1)e^u r\frac{\partial u}{\partial r}$. Since

$$\begin{aligned}
\int_{B_1} r\tfrac{\partial v}{\partial r}\, dx &= \int_{B_1} x \bullet \nabla v\, dx \\
&= \int_{\partial B_1} v\tfrac{\partial}{\partial \eta}[\nabla(\tfrac{1}{2}r^2)]\, ds - \int_{B_1} v\Delta(\tfrac{1}{2}r^2)\, dx \\
&= -2\delta \int_{\partial B_1} r\tfrac{\partial r}{\partial \eta}\, ds - \int_{B_1} nv\, dx \\
&= -2\delta w_n - \int_{B_1} nv\, dx,
\end{aligned}$$

we have $\int_{B_1}(r\frac{\partial v}{\partial r} + nv)\, dx = -2\delta w_n$ where w_n is the surface area of B_1.

Define $I := \int_{B_1}[-u\Delta(r\frac{\partial u}{\partial r}) + r\frac{\partial u}{\partial r}\Delta u]\, dx$. Then by (2.25),

$$\begin{aligned}
I &= \int_{B_1}[\delta u e^u(r\tfrac{\partial u}{\partial r} + 2) - \delta e^u r\tfrac{\partial u}{\partial r}]\, dx \\
&= \int_{B_1}[\delta(u-1)e^u r\tfrac{\partial u}{\partial r} + 2\delta u e^u]\, dx \\
&= \int_{B_1}[r\tfrac{\partial v}{\partial r} + 2v + 4\delta e^u]\, dx \\
&= \int_{B_1}[r\tfrac{\partial v}{\partial r} + nv + (2-n)v + 4\delta e^u]\, dx \\
&= -2\delta w_n + \int_{B_1}[(2-n)u + 2n]\delta e^u\, dx.
\end{aligned}$$

But by Green's identity,

$$I = \int_{\partial B_1}\left[-u\frac{\partial}{\partial \eta}\left(r\frac{\partial u}{\partial \eta}\right) + r\frac{\partial u}{\partial r}\frac{\partial u}{\partial \eta}\right] ds = \int_{\partial B_1} r\left(\frac{\partial u}{\partial \eta}\right)^2 ds$$

since $u = 0$ on ∂B_1, $r = 1$, and $\frac{\partial}{\partial \eta} = \frac{\partial}{\partial r}$ on ∂B_1; so $I = \beta^2 w_n$. Thus,

$$\int_{B_1}[(2-n)u + 2n]\delta e^u\, dx = (\beta^2 + 2\delta)w_n \qquad (2.26)$$

which completes the proof. □

Lemma 2.23 *For $n \geq 3$, if $(\delta, \beta) \in \mathcal{D}$, then $\beta^2 - 2n\beta + 2\delta < 0$.*

Proof. By Green's identity, $\int_{B_1} \delta e^u \, dx = \beta w_n$. Thus, (2.26) gives us

$$\int_{B_1} \delta u e^u \, dx = \frac{\beta^2 - 2n\beta + 2\delta}{2-n} w_n.$$

Since the integral is positive and since $n \geq 3$, we necessarily have $\beta^2 - 2n\beta + 2\delta < 0$. □

Lemma 2.24 *The bifurcation curve $\mathcal{D}$ intersects the graph of $\beta = L(\delta) = \frac{\delta}{n-1}$ in at least one point where $n - 1 < \delta < 2(n-1)$.*

Proof. Let $\beta(\delta)$ be the arc of $\mathcal{D}$ which originates at $(0, 0)$ and terminates at $(\delta_{FK}, 2)$. From (2.22) we obtain

$$\beta'(\delta) = \frac{(n-2)\beta - \delta}{\delta(\beta - 2)} \tag{2.27}$$

By Lemma 2.23, for $\beta = 2$, we have $\delta < 2(n-1)$ and so $\beta(\delta)$ reaches $\beta = 2$ at $\delta < 2(n-1)$. Observe that $\beta'(0) = \frac{1}{n} < \frac{1}{n-1} = L'(0)$. Since $L(2(n-1)) = 2$, $\beta(\delta)$ intersects $L(\delta)$ at a value $\delta < 2(n-1)$.

If $\beta(\delta_0) = L(\delta_0) = \beta_0$ for $\beta_0 \in (0, 1]$, that is, for $\delta_0 \in (0, n-1]$, then $\beta'(\delta_0) = [(n-1)(2-\beta_0)]^{-1} < (n-1)^{-1} = L'(\delta_0)$. Thus, if there are any points of intersection for $\beta_0 \in (0, 1]$, then there is only one. This implies $\beta'(0) \geq (n-1)^{-1}$. But $\beta'(0) = n^{-1} < (n-1)^{-1}$, so there are no points of intersection for $\beta \leq 1$.

Since $\beta'(\delta) = [(n-1)(2-\beta)]^{-1} > (n-1)^{-1}$ for any (δ, β) with $\beta = L(\delta)$ and $n - 1 < \delta < 2(n-1)$, the intersection is unique on that arc. □

For $n \geq 10$, the shape of the bifurcation curve $\mathcal{D}$ guarantees that $\mathcal{D}$ and the line $\beta = L(\delta)$ intersect at a unique point.

Lemma 2.25 *For $3 \leq n \leq 9$, $\mathcal{D}$ intersects $L = \{(\delta, \beta) : \delta > 0, \beta = \frac{\delta}{n-1}\}$ uniquely.*

Proof. By Lemma 2.24, $\mathcal{D} \cap L \neq \emptyset$. We claim that there are no other intersections as $\mathcal{D}$ spirals towards $(2(n-2), 2)$.

Let R be the region bounded by

$$\begin{aligned}
L_1 &= \{(\delta, \beta) : \delta = 2(n-1),\ 2 \leq \beta \leq 2\tfrac{n-1}{n-2}\}, \\
L_2 &= \{(\delta, \beta) : \beta = 2\tfrac{n-1}{n-2},\ n \leq \delta \leq 2(n-1)\}, \\
L_3 &= \{(\delta, \beta) : \beta = -\tfrac{1}{4}\delta^2 + \tfrac{3}{2}\delta + \tfrac{3}{4},\ 1 \leq \delta \leq 3\} \text{ for } n = 3, \\
&= \{(\delta, \beta) : \beta = \tfrac{\delta}{n-1} + 1,\ n-2 \leq \delta \leq n\} \text{ for } n \geq 4, \\
L_4 &= \{(\delta, \beta) : \delta = n-2,\ 1 \leq \beta \leq 2\}, \\
L_5 &= \{(\delta, \beta) : \beta = 1,\ n-2 \leq \delta \leq n-1\}, \text{ and} \\
L_6 &= \{(\delta, \beta) : \beta = \tfrac{\delta}{n-1},\ n-1 \leq \delta \leq 2(n-1)\}.
\end{aligned}$$

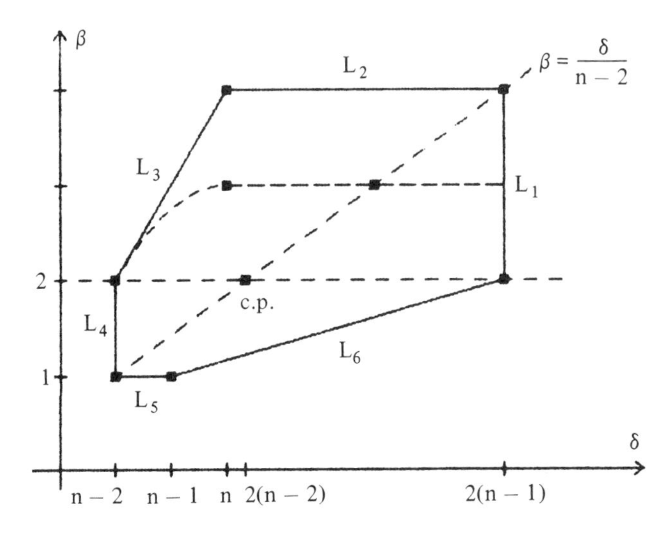

Figure 2.9.

Figure 2.9 illustrates the region R.
Observe that $\beta'(\delta) < 0$ on

$$\begin{aligned} S &= \{(\delta,\beta) : \delta > 2(n-2), 2 < \beta < \tfrac{\delta}{n-2}\} \\ &\quad \cup\{(\delta,\beta) : 0 < \delta < 2(n-1), \tfrac{\delta}{n-2} < \beta < 2\} \end{aligned}$$

and $\beta'(\delta) > 0$ on $\{(\delta,\beta) : \beta > 0, \delta > 0\} \setminus S$. Thus, $\mathcal{D}$ cannot leave R through L_1 or L_2. The orbit cannot leave through L_3, L_4, L_5, or L_6 since the slope at such a crossing would not agree with $\beta'(\delta)$ evaluated on these sets. Thus, the first point of intersection of $\mathcal{D}$ with L_6 is the only point of intersection. □

This completes the proof of (i) for $n \geq 2$. We now show that points of inflection for the graph of $u(r)$ are unique, using the fact that $\mathcal{D}$ intersects $L(\delta)$ uniquely.

Lemma 2.26 *Consider (2.17)-(2.18) with $(n-1)\beta - \delta = 0$ for $n \geq 2$. There exists one and only one solution $u(r)$ with $\delta = (n-1)\beta$. This guarantees a unique solution such that $u''(1) = 0$.*

Proof. For $n \geq 3$, $\mathcal{D}$ intersects L at the unique point $(\overline{\delta}, \overline{\beta})$. This gives the unique solution $u(r)$. For $n = 2$, $\mathcal{D}$ is given by $\beta^2 - 4\beta + 2\delta = 0$ which intersects $\beta - \delta = 0$ uniquely at $(\overline{\delta}, \overline{\beta}) = (2, 2)$. □

The proof of Theorem 2.21 will be complete if we can show any solution $u(r)$ of (2.17) has at most one inflection point for $n \geq 2$.

Lemma 2.27 *Let $u(r) \in C^2([0,1],\mathbb{R})$ be a solution of (2.17) for $n \geq 2$; then u has at most one inflection point. In fact, if $u''(1) > 0$, then $u'' = 0$ for a unique $r \in (0,1)$. If $u''(1) < 0$, then u is concave on $[0,1]$.*

Proof. Let $R \in (0,1)$ be the first value of r such that $u''(r) = 0$. Define $m := u'(R)$; then $u(R) = \ln\left[\frac{-m(n-1)}{\delta R}\right]$. In (2.17), let $r = sR$ and $v(s) = u(r) - u(R)$. For $s \in [0,1]$, we have

$$\begin{aligned} & v'' + \tfrac{n-1}{s}v' + \overline{\delta}e^v = 0, \ 0 < s < 1 \\ & v'(0) = 0, \ v(1) = 0, \ v'(1) = -\overline{\beta} \qquad (2.28) \\ & (n-1)\overline{\beta} - \overline{\delta} = 0 \end{aligned}$$

where $\overline{\delta} = -(n-1)mR > 0$ and $\overline{\beta} = -mR > 0$.

By Lemma 2.26, there is a unique pair $(\overline{\delta}, \overline{\beta})$ satisfying $(n-1)\overline{\beta} - \overline{\delta} = 0$ and a corresponding unique solution $v(s)$. Thus, $v''(1) = 0$. Since $u''(r) < 0$ on $0 \leq r < R$, $v''(s) < 0$ for $0 \leq s < 1$.

Suppose there as a value $P \in (R,1]$ such that $u''(R) = u''(P) = 0$. Set $\ell = u'(P)$; then $u(P) = \ln\left[\frac{-\ell(n-1)}{\delta P}\right]$. Make a change of variables $r = sP$ and $v(s) = u(r) - u(P)$. Restricting $s \in [0,1]$ we have that $v(s)$ satisfies (2.28) with $\overline{\delta} = -(n-1)\ell P$ and $\overline{\beta} = -\ell P > 0$. By uniqueness, $v''(s) < 0$ for $s \in [0,1]$. But

$$v''\left(\frac{R}{P}\right) = P^2 u''(R) = 0$$

with $0 < \frac{R}{P} < 1$ is a contradiction. □

These last lemmas prove (ii), so the proof of Theorem 2.21 is complete. □

We now turn our attention to the perturbed Gelfand problem:

$$\begin{aligned} & -\Delta u = \delta \exp\left(\tfrac{u}{1+\varepsilon u}\right), \ \ x \in B_1 \\ & u(x) = 0, \ \ x \in \partial B_1 \end{aligned} \qquad (2.29)$$

where $\delta > 0$ and $\varepsilon > 0$. By the maximum principle, all solutions to (2.29) are positive on B_1. By Theorem 2.17, all solutions must be radially symmetric. BVP (2.29) is equivalent to

$$\begin{aligned} & u'' + \tfrac{n-1}{r}u' + \delta \exp\left(\tfrac{u}{1+\varepsilon u}\right) = 0, \ \ 0 < r < 1 \\ & u'(0) = 0, \ u(1) = 0. \end{aligned} \qquad (2.30)$$

We use the same notation as before: $\alpha = u(0)$ and $\beta = -u'(1)$. For ε sufficiently close to 0, we can obtain limited information about solution profiles for (2.30). This information is not as precise as that for the Gelfand problem, but it appears that such precision is attainable with more detailed work. (See Figure 2.8 and the comments following it.)

Theorem 2.28 *Consider BVP (2.30).*

1. *For $n = 1$, $\varepsilon > 0$, and $\delta > 0$, every solution is concave.*

2. *For $n = 2$, all solutions are bell-shaped or concave on $[0, 1)$. The solutions corresponding to $(\delta, \beta) \in \mathcal{D}_\varepsilon$ (the bifurcation curve) with $\beta < \delta$ are concave.*

3. *For $n \geq 3$ and $\varepsilon > 0$ sufficiently small, there are values $\delta_1(\varepsilon)$ and $\delta_2(\varepsilon)$ such that the minimal solution is concave down for $0 < \delta < \delta_1$ and not concave for $\delta_2 < \delta < \delta_{FK}(\varepsilon)$.*

The results are proved by the following set of lemmas.

Lemma 2.29 *For $\varepsilon > 0$ and sufficiently close to zero, any large solution of (2.30) must satisfy $u''(1) > 0$.*

Proof. We point out that, as in the results for the Gelfand problem, $\text{sgn}[u''(1)] = \text{sgn}[(n-1)\beta - \delta]$. Let $u(r)$ be a large solution to (2.30). Let $\mu \in (0, 1)$. There exists a constant k such that $u(r) \geq ke^{1/\varepsilon}$ for $r \in [0, \mu]$. Let $f(u) = exp(\frac{u}{1+\varepsilon u})$. Integration of the differential equation in (2.30) yields

$$\begin{aligned}\beta &= \delta \int_0^1 r^{n-1} f(u(r))dr \geq \delta \int_0^\mu r^{n-1} f(u(r))dr \\ &\geq \delta \int_0^\mu r^{n-1} f(ke^{1/\varepsilon})dr = \delta \tfrac{\mu^n}{n} f(ke^{1/\varepsilon}) > \tfrac{\delta}{n-1}\end{aligned}$$

for ε sufficiently close to 0. Thus, $u''(1) = (n-1)\beta - \delta > 0$ for ε sufficiently close to 0. □

Lemma 2.30 *Let $n = 2$. Points of inflection to solutions of (2.30) are unique. Consequently, all solutions are either bell-shaped or concave down.*

Proof. Differentiating in (2.30) gives us

$$u''' + \left(\frac{ru'' - u'}{r}\right) + \delta f'(u)u' = 0.$$

Suppose $(R, u(R))$ is a point of inflection. Then

$$u'''(R) = \frac{u'(R)}{R^2}\left(1 + \frac{Ru'(R)}{[1+\varepsilon u(R)]^2}\right) \tag{2.31}$$

where use has been made of $u''(R) = 0$. The function $ru'(r)$ is decreasing and the function $[1 + \varepsilon u(r)]^{-2}$ is increasing. Suppose $(P, u(P))$ is another point of inflection where $R < P < 1$, $u''(r) < 0$ for $r \in [0, R)$, and $u''(r) > 0$ for $R < r < P$; then

$$1 + \frac{Pu'(P)}{[1+\varepsilon u(P)]^2} \leq 1 + \frac{Ru'(R)}{[1+\varepsilon u(R)]^2} < 0$$

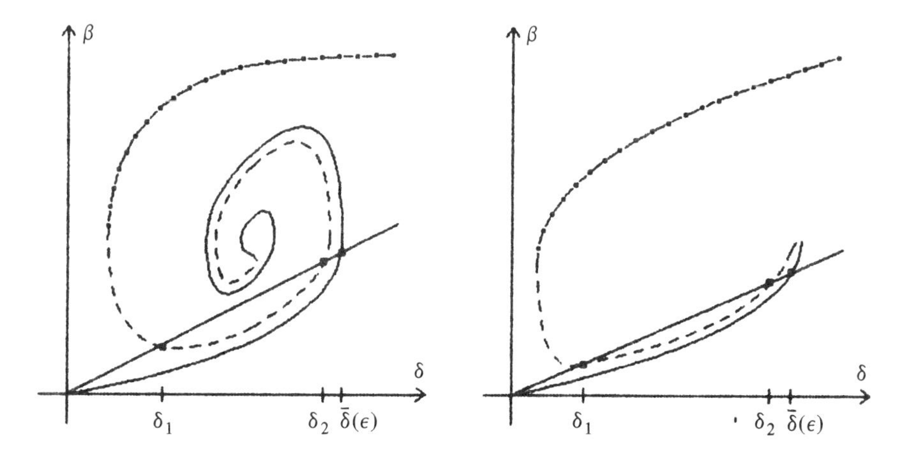

Figure 2.10.

where the last inequality is valid since $u'''(R) > 0$. This set of inequalities forces $u'''(P) > 0$ which cannot happen, so $(R, u(R))$ is the only point of inflection. As a consequence, a solution to (2.30) is either concave down or bell-shaped on $[0, 1)$. □

Lemma 2.31 *Let $n \geq 3$. For $\varepsilon > 0$ sufficiently close to 0, the graphs of the bifurcation curve and the line $(n-1)\beta - \delta = 0$ intersect in at least two points with the large–small branch and in a unique point with the small–small branch at the point $(\overline{\delta}(\varepsilon), \overline{\beta}(\varepsilon))$.*

Proof. From the results on the Gelfand problem, we had a unique point of intersection $(\delta, \beta) = (\overline{\delta}(0), \overline{\beta}(0))$ where the function notation indicates $\varepsilon = 0$. At this point of intersection, the angle of intersection is positive so that a small perturbation of the bifurcation curve still yields a unique point of intersection $(\overline{\delta}(\varepsilon), \overline{\beta}(\varepsilon))$. By our earlier remark on the closeness of the large–small branch to the small–small branch, the large–small branch and the straight line must intersect in at least two more points. For ε sufficiently close to zero, one of these points of intersection must occur near $(\overline{\delta}(\varepsilon), \overline{\beta}(\varepsilon))$, the other near $(0, 0)$. Figure 2.10 illustrates this.

At these points of intersection, $u''(1) = 0$. From (2.31), $u'''(1) = -\beta(1-\beta)$. Thus, at (δ_1, β_1) near $(0, 0)$, $u'''(1) < 0$ and at (δ_2, β_2), $(\overline{\delta}(\varepsilon), \overline{\beta}(\varepsilon))$ near $(\overline{\delta}(0), \overline{\beta}(0))$, $u'''(1) > 0$ for solutions $u(r)$ corresponding to these pairs of (δ, β).

The solution corresponding to the pair (δ_1, β_1) consequently must have at least two points of inflection so that the graph of $u(r)$ may reach the u-axis. □

2.5 Comments

In this chapter we have given a detailed discussion of the steady-state models and their natural generalizations. We have referred to the steady-state model (1.30)-(1.31) as the Gelfand problem and considered the radially symmetric case (2.16). The small fuel loss model (2.23) was referred to as the perturbed Gelfand problem. Gelfand [GEL] appears to have been the first to make an indepth study of (1.30)-(1.31) although the problem has a long history in the radially symmetric case in low dimensions.

For dimension $n = 1$, Liouville [LIU] first studied and found an explicit solution in 1853. For $n = 2$, Bratu [BRA] found an explicit solution in 1914. Frank-Kamenetski [FRA] rediscovered these results in his development of thermal explosion theory. Joseph and Lundgren [JOS] gave an elementary proof via a phase plane analysis of the multiple existence of solutions for dimensions $n \geq 3$. These results are summarized as Theorem 2.19.

The idea of using upper and lower solutions to establish existential results for nonlinear boundary value problems goes back to Nagumo [NAG1], [NAG3], [NAG5]. There are several excellent papers [SCH],[LIN], [SAT1] which survey boundary value problems of the type (2.1) and (2.2). Although very general existential results are known, the problem of multiplicity of solutions remains open for arbitrary domains, even for the Gelfand problem.

The radial symmetry results of Section 2.2 are due to Gidas, Ni, and Nirenberg [GID1]. It is remarkable that such a result was not rigorously proved until the late 1970's. Troy [TRO1] generalized these results to systems where the nonlinearity is quasimonotone.

The results in Section 2.4 on the shape of solutions for the Gelfand problem and the perturbed Gelfand problem are due to the authors [BEB8]. These results show that (2.24) and (2.29) exhibit a "hot spot" development. The results for the perturbed Gelfand problem are not as precise as those for the Gelfand problem. The problem of determining the exact qualitative shapes of solutions to the perturbed Gelfand problem is an open question.

3
The Rigid Ignition Model

We wish to analyze indepth the solid fuel ignition model (1.28)-(1.29)

$$\theta_t - \Delta\theta = \delta e^{\theta}, \quad (x,t) \in \Omega \times (0,T)$$
$$\theta(x,0) = 0, \quad x \in \Omega$$
$$\theta(x,t) = 0, \quad (x,t) \in \partial\Omega \times (0,T)$$

and its relationship to the steady-state model (1.30)-(1.31)

$$-\Delta\psi = \delta e^{\psi}, \quad x \in \Omega$$
$$\psi(x) = 0, \quad x \in \partial\Omega.$$

Existence-uniqueness for (1.28)-(1.29) is established for a more general initial-boundary value problem with nonlinearity $f(x,u)$ and with initial data $\phi(x)$ not necessarily 0. The results use the ideas of upper and lower solutions, invariance, and comparison. Other properties of solutions to (1.28)-(1.29) are also determined by application of these ideas.

For nonlinearities of the type $f(u) = e^u$ (or $f(u) = u^p$), the solutions to (1.28)-(1.29) do not exist for all $t \in (0,\infty)$, and the solutions become unbounded at a first time $T < \infty$, called the blowup time for the problem. In Section 3.2, we determine if blowup occurs, and if so, then when does it occur; that is, we determine upper bounds on the blowup time. If the solutions do exist for all time, then for appropriate initial data the solutions to (1.28)-(1.29) converge to a solution of (1.30)- (1.31). The critical value δ_{FK} seems to provide a separation between blowup of solutions ($\delta > \delta_{FK}$) and global existence ($\delta < \delta_{FK}$).

We also consider the question of where blowup occurs. The main result of Section 3.3 is Theorem 3.16 which guarantees under certain restrictions on the nonlinearity that blowup occurs only at a single point (for radial domains). Other information obtained in the proofs lead to bounds on the solutions for spatial values near the blowup point.

The majority of the chapter is contained in Section 3.4. We consider radially symmetric solutions $u(r,t)$ on a ball and transform the problem (1.28)-(1.29), for $f(u) = e^u$ or $f(u) = u^p$, using the idea of self- similarity. We analyze the solutions u for which blowup occurs and determine their profiles near a blowup singularity in an asymptotic sense in space-time parabolas containing the blowup point. One major difficulty is that of establishing *a priori* bounds on the solution u and its derivatives which allow us to choose the correct steady-state solution (in the self-similar sense) which u

converges to. Another difficulty is in showing that u does converge to a steady-state solution which is independent of the self-similar time variable. Energy integral estimates are used in proving this convergence.

Consequences of the analysis in Section 3.4 are the following: For $f(u) = e^u$, the solution u satisfies $u(0,t) \sim -\ln(T-t)$ as $t \to T^-$ (for blowup time T). For $f(u) = u^p (p > 1)$, the solution u satisfies $u(0,t) \sim [\beta(T-t)]^\beta$ as $t \to T^-$ where $\beta = \frac{1}{p-1}$.

3.1 Existence-Uniqueness

Let $\Omega \subset \mathbb{R}^n$ be a bounded domain whose boundary $\partial\Omega$ is an $n-1$ dimensional manifold of class $C^{2+\alpha}$ for some $0 < \alpha < 1$. Let $\Pi_T = \Omega \times (0,T)$ and $\Gamma_T = [\partial\Omega \times (0,T)] \cup [\overline{\Omega} \times \{0\}]$. Assume that $f : \overline{\Omega} \times [0,T] \times \mathbb{R} \to \mathbb{R}$ is a locally Hölder continuous function with Hölder exponents α, $\frac{1}{2}\alpha$, and α in the respective variables x, t, and u. Assume that $\psi : \Gamma_T \to \mathbb{R}$ is continuous. Consider the partial differential equation

$$u_t - \Delta u = f(x,t,u), \quad (x,t) \in \Pi_T \tag{3.1}$$

with initial-boundary condition

$$u(x,t) = \psi(x,t), \quad (x,t) \in \Gamma_T. \tag{3.2}$$

Definition 3.1 *A function $v \in C(\overline{\Pi}_T, \mathbb{R}) \cap C^{2,1}(\Pi_T, \mathbb{R})$ is* lower solution *of (3.1)-(3.2) if*

$$\begin{aligned} v_t - \Delta v &\le f(x,t,v), \quad (x,t) \in \Pi_T \\ v(x,t) &\le \psi(x,t), \quad (x,t) \in \Gamma_T. \end{aligned}$$

If the inequalites above are reversed, then v is called an upper solution *of (3.1)-(3.2).*

The following theorem is a consequence of invariance and will be proved in Chapter 4.

Theorem 3.1 *Let α be a lower solution and let β be an upper solution of IBVP (3.1)-(3.2) with $\alpha(x,t) \le \beta(x,t)$ on $\overline{\Pi}_T$; then (3.1)-(3.2) has a solution $u \in C^{2,1}(\Pi_T)$ with $\alpha(x,t) \le u(x,t) \le \beta(x,t)$ on $\overline{\Pi}_T$.*

The next theorem will also only be stated at this time. Its proof will follow immediately from our comparison theorems (Theorem 4.1), Corollary 4.2) given in Chapter 4.

Theorem 3.2 *Let $u, v \in C(\overline{\Pi}_T, \mathbb{R}) \cap C^{2,1}(\Pi_T, \mathbb{R})$ with*

1. *$u_t - \Delta u - f(x,t,u) < v_t - \Delta v - f(x,t,v)$ for $(x,t) \in \Pi_T$, and*

2. $u(x,t) < v(x,t)$ *for* $(x,t) \in \Gamma_T$*;*

then $u(x,t) < v(x,t)$ *for* $(x,t) \in \overline{\Pi}_T$. *In addition, if* f *is locally Lipschitz continuous in* u*, then the result is true with weak inequalities.*

We now consider a special case of initial-boundary value problem (3.1)-(3.2) where the function f is independent of t and is Lipschitz continuous in u:

$$u_t - \Delta u = f(x,u), \quad (x,t) \in \Pi_T \tag{3.3}$$

with initial-boundary conditions

$$\begin{aligned} u(x,0) &= \phi(x), \quad x \in \Omega \\ u(x,t) &= 0, \quad (x,t) \in \partial\Omega \times (0,\infty). \end{aligned} \tag{3.4}$$

Theorem 3.3 *Let* $u(x,t)$ *be a solution of IBVP (3.3)-(3.4). If* $\phi(x)$ *is a lower solution of (3.3), then* $u(x,t)$ *is nondecreasing in* t *for each fixed* x.

Proof. By Theorem 3.2, $u(x,t) \geq \phi(x)$. For $\delta > 0$, define $u^\delta(x,t) = u(x,t+\delta)$; then u^δ is a solution of (3.3) and so is an upper solution with $u^\delta(x,0) = u(x,\delta) \geq \phi(x) = u(x,0)$. Thus, $u^\delta(x,t) \geq u(x,t)$ for all $(x,t) \in \overline{\Pi}_T$ and so $u(x,t)$ is nondecreasing in t. □

Theorem 3.4 *Let* $\alpha(x)$ *be a bounded lower solution and let* $\beta(x)$ *be a bounded upper solution to IBVP (3.3)-(3.4) with* $\alpha(x) < \beta(x)$ *on* Ω*; then (3.3)-(3.4) has a unique solution* $u(x,t)$ *with*

$$\alpha(x) \leq u(x,t) \leq \beta(x) \quad \text{and} \quad \lim_{t\to\infty} u(x,t) = u_0(x)$$

where the limit is uniform in x *and where* $u_0(x)$ *is the minimal solution of*

$$\begin{aligned} -\Delta u_0 &= f(x,u_0), \quad x \in \Omega \\ u_0(x) &= 0, \quad x \in \partial\Omega. \end{aligned}$$

Proof. By Theorems 3.1 and 3.2, (3.3)-(3.4) has a unique solution $u(x,t)$ with $\alpha(x) \leq u(x,t) \leq \beta(x)$. By Theorem 3.3, $u(x,t)$ is nondecreasing in t for each fixed x. In addition, $u(x,t)$ is bounded above, so $\lim_{t\to\infty} u(x,t) = u_0(x)$ pointwise for each $x \in \Omega$. By Dini's Theorem, the convergence is uniform on compact subsets of Ω.

Set $u_n(x,t) = u(x,t+n)$ for $t \in [0,1]$ and $n \in \mathbb{N}$; then $u_n(x,t)$ is the solution of

$$\begin{aligned} &w_t - \Delta w = f(x, u_n(x,t)), \quad (x,t) \in \Omega \times (0,\infty) \\ &w(x,0) = u_n(x,0), \quad x \in \overline{\Omega} \\ &w(x,t) = 0, \quad (x,t) \in \partial\Omega \times (0,\infty). \end{aligned}$$

For the appropriate Green's function $G(x,t)$ we have

$$u_n(x,t) = \int_\Omega G(x-y,t)u_n(y,0)\,dy + \int_0^t d\tau \int_\Omega G(x-y,t-\tau)f(y,u_n)\,dy$$

for $0 \leq t \leq 1$. By the Lebesgue Dominated Convergence Theorem, letting $n \to \infty$, we have

$$\begin{aligned} u_0(x) &= \textstyle\int_\Omega G(x-y,t)u_0(y)\,dy \\ &\quad + \textstyle\int_0^t d\tau \int_\Omega G(x-y,t-\tau)f(y,u_0(y))\,dy \\ &=: I_1(x,t) + I_2(x,t) \end{aligned}$$

with $u_0(x)$ continuous.

Since $I_1(x,t)$ and $I_2(x,t)$ are differentiable with respect to x, $u_0(x)$ is differentiable. Since f is locally Hölder in x and u, we have $u_0(x)$ is twice-differentiable in x, and: $-\Delta u_0 = f(x,u_0)$ for $x \in \Omega$ and $u_0(x) = 0$ for $x \in \partial\Omega$. □

Theorem 3.5 *For any $\delta > 0$, there exists $T > 1/\delta$ such that IBVP (1.28)-(1.29) has a unique solution $\theta(x,t)$ on $\overline{\Omega} \times [0,T)$ with*

$$0 \leq \theta(x,t) \leq -\ln(1-\delta t)$$

for $(x,t) \in \overline{\Omega} \times [0,1/\delta)$.

Proof. Set $\alpha(x,t) = 0$. Set $\beta(t) = -\ln(1-\delta t)$, which is the solution of

$$\beta' = \delta e^\beta, \quad \beta(0) = 0.$$

Since α is a lower solution and β is an upper solution of IBVP (1.28)-(1.29) with $\alpha < \beta$, we conclude by Theorems 3.1 and 3.2 that IBVP (1.28)-(1.29) has a unique solution $\theta(x,t)$ on $\overline{\Omega} \times [0,t^*)$, $t^* \geq 1/\delta$, with

$$0 \leq \theta(x,t) \leq -\ln(1-\delta t)$$

for $(x,t) \in \overline{\Omega} \times [0,1/\delta)$. □

The inequality for $\theta(x,t)$ in Theorem 3.5 is illustrated in Figure 3.1.

Theorem 3.6 *For $\delta < \delta_{FK}$, IBVP (1.28)-(1.29) has a unique solution $\theta(x,t)$ on $\overline{\Omega} \times [0,\infty)$ with $0 \leq \theta(x,t) \leq \phi(x)$ where $\phi(x)$ is the minimal solution of BVP (1.30)-(1.31).*

Proof. Set $\alpha(x,t) = 0$. Set $\beta(x,t) = \phi(x)$ where ϕ is the minimal solution of BVP (1.30)-(1.31) (which exists for $\delta < \delta_{FK}$). By Theorems 3.1 and 3.2, IBVP (1.28)- (1.29) has a unique solution $\theta(x,t)$ on $\overline{\Omega} \times [0,\infty)$ with

$$0 \leq \theta(x,t) \leq \phi(x)$$

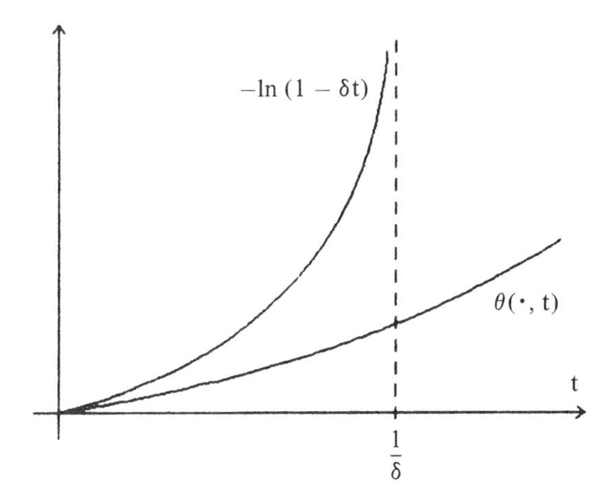

Figure 3.1.

for $(x,t) \in \overline{\Omega} \times [0,\infty)$, and by Theorems 3.3 and 3.4,

$$\lim_{t\to\infty} \theta(x,t) = \phi(x)$$

uniformly in x. □

For BVP (1.30)-(1.31) we know there is a $\delta_{FK} > 0$ such that no solution exists for $\delta > \delta_{FK}$. We now ask what happens to IBVP (1.28)-(1.29) for $\delta > \delta_{FK}$.

Theorem 3.7 *For each $\delta > \delta_{FK}$, there is a $T \in [1/\delta, \infty)$ such that IBVP (1.28)-(1.29) has a unique solution $\theta(x,t)$ on $\overline{\Omega} \times [0,T)$. Moreover,*

$$\lim_{t\to T^-} \max\{\theta(x,t) : x \in \overline{\Omega}\} = \infty.$$

Proof. For each $n \in \mathbb{N}$, define $f_n(y) = \min\{\delta e^y, \delta e^n\}$ and consider

$$y_t - \Delta y = f_n(y), \quad (x,t) \in \Pi = \Omega \times (0,\infty) \tag{3.5}$$

$$y(x,t) = 0, \quad (x,t) \in \Gamma = [\partial\Omega \times (0,\infty)] \cup [\overline{\Omega} \times \{0\}]. \tag{3.6}$$

By constructing a sequence of solutions to an associated sequence of nonhomogeneous linear initial-boundary value problems, we shall prove that IBVP (3.5)-(3.6) has a unique solution $u_n(x,t)$ on $\overline{\Omega} \times [0,\infty)$.

Set $u^0(x,t) = 0$ on $\overline{\Pi}$ and let $u^1(x,t)$ be the unique solution of the nonhomogeneous linear problem:

$$u_t^1 - \Delta u^1 = f_n(u^0(x,t)), \quad (x,t) \in \Pi$$

$$u^1(x,t) = 0, \quad (x,t) \in \Gamma.$$

Since

$$u_t^1 - \Delta u^1 = \delta e^0 = \delta > 0 = u_t^0 - \Delta u^0, \quad (x,t) \in \Pi,$$
$$u^1(x,t) = u^0(x,t), \quad (x,t) \in \Gamma,$$

Theorem 3.2 implies that $u^1 \geq u^0$ on $\overline{\Pi}$.

Let $u^2(x,t)$ be the unique solution of:

$$u_t^2 - \Delta u^2 = f_n(u^1(x,t)), \quad (x,t) \in \Pi,$$
$$u^2(x,t) = 0, \quad (x,t) \in \Gamma.$$

Since

$$u_t^2 - \Delta u^2 = f_n(u^1) \geq f_n(u^0) = u_t^1 - \Delta u^1, \quad (x,t) \in \Pi,$$
$$u^2(x,t) = u^1(x,t), \quad (x,t) \in \Gamma,$$

we have $u^2 \geq u^1$ on $\overline{\Pi}$.

Continuing in this way, we construct a sequence $\{u^k\}$ such that for each $k \geq 1$, u^k is the unique solution of:

$$u_t^k - \Delta u^k = f_n(u^{k-1}), \quad (x,t) \in \Pi,$$
$$u^k(x,t) = 0, \quad (x,t) \in \Gamma.$$

Also,

$$0 \leq u^0 \leq u^1 \leq \cdots \leq u^k \leq \cdots$$

for $\overline{\Pi}$, and by Theorem 3.2,

$$u^k \leq -\ln(1 - \delta t)$$

for $(x,t) \in \overline{\Omega} \times [0, (1 - e^{-n})/\delta)$ for all $k \geq 0$.

Since $\{f_n(u^{k-1})\}$ is uniformly bounded, $\{u^k\}$ is bounded above. Hence, $u^k(x,t) \to u_n(x,t)$ pointwise on $\overline{\Pi}$. By standard bootstrapping arguments we have that $u_n(x,t)$ is the solution of: $u_t - \Delta u = f_n(u)$ for $(x,t) \in \Pi$ and $u(x,t) = 0$ for $(x,t) \in \Gamma$.

On $\Omega \times [0, (1 - e^{-n})/\delta)$, $u_n(x,t) = \theta(x,t)$ where θ is the unique solution of IBVP (1.28)-(1.29). If $u_n(x,t) \leq n$ for all $(x,t) \in \overline{\Omega} \times [0, \infty)$, then $u_n(x,t) \to \phi(x)$ uniformly in x as $t \to \infty$ where ϕ is the minimal solution of BVP (1.30)-(1.31). But $\delta > \delta_{FK}$, so no such solution exists. There must exist $(x_n, t_n) \in \Pi$ such that $u_n(x_n, t_n) = n$ and $u_n(x,t) = \theta(x,t)$ where θ is the unique solution to BVP (1.30)-(1.31) on $\overline{\Omega} \times [0, t_n)$ for each $n \in \mathbb{N}$. By compactness of $\overline{\Omega}$, there exists a subsequence $\{(x_{n_k}, t_{n_k})\}$ of $\{(x_n, t_n)\}$ such that

$$x_{n_k} \to x^* \in \overline{\Omega} \text{ and } t_{n_k} \uparrow T \leq \infty \text{ as } k \to \infty$$

with $u_{n_k}(x,t) = \theta(x,t)$ on $\overline{\Omega} \times [0, t_{n_k})$. Thus,

$$\lim_{t \to T^-} \max_{x \in \overline{\Omega}} \theta(x,t) = \infty,$$

and the proof is complete. □

For the equation

$$u_t - \Delta u = \delta f(u), \quad (x,t) \in \Omega \times [0,\infty) \tag{3.7}$$

with initial-boundary conditions

$$\begin{aligned} &u(x,0) = u_0(x), \quad x \in \Omega \\ &\tfrac{\partial u}{\partial \eta} + \beta u = 0, \quad (x,t) \in \partial\Omega \times (0,\infty), \quad 0 < \beta(x) \leq \infty \end{aligned} \tag{3.8}$$

where $f(u) \geq 0$, $f'(u) \geq 0$, and $f''(u) \geq 0$ for $u \geq 0$, similar results hold. For the boundary value problem

$$-\Delta u = \delta f(u), \quad x \in \Omega \tag{3.9}$$

$$\frac{\partial u}{\partial \eta} + \beta u = 0, \quad x \in \partial\Omega \tag{3.10}$$

there is a $\delta_{FK} > 0$ such that solutions exist for $0 < \delta < \delta_{FK}$ and no solutions exist for $\delta > \delta_{FK}$.

Thus, if $\delta < \delta_{FK}$ and if w_m is the minimal solution of (3.9)-(3.10) and if $u_0 \leq w_m$, then taking a lower solution $\alpha \leq u_0 \leq w_m$, we have $u(x,t) \to w_m$ as $t \to \infty$ and $u(x,t)$ exists globally. If $\delta > \delta_{FK}$, the solution $u(x,t)$ becomes unbounded in the L^∞-sense as $t \to T^-$.

3.2 Blowup: When?

In the last section we discussed existence of solutions to certain initial-boundary value problems. This section deals with the determination of the maximum time interval for which solutions exist.

Definition 3.2 *The solution $u(x,t)$ of IBVP (3.7)-(3.8) for $\delta > \delta_{FK}$ becomes unbounded as $t \to T_-$. We say that* thermal runaway *or* blowup *occurs at T.*

For IBVP (1.28)-(1.29), this says that the thermal event is explosive (supercritical).

By the analogue of Theorem 3.7 for IBVP (3.7)-(3.8), we cannot determine if blowup occurs in finite or infinite time. A necessary condition for blowup in finite time is the following.

Theorem 3.8 *If the unique solution $u(x,t)$ of IBVP (3.7)-(3.8) blows up in finite time T, then*

$$\int_b^\infty [f(s)]^{-1}\, ds < \infty \text{ for } b \geq 0. \tag{3.11}$$

Proof. Assume that $\int^\infty [f(s)]^{-1}\, ds = \infty$. Let $\beta(t)$ be the solution of $u' = \delta f(u)$ with $u(0) = \sup\{u_0(x) : x \in \overline{\Omega}\}$; then $\beta(t)$ is an upper solution of (3.7)-(3.8) and so $\beta(t) \geq u(x,t) \geq 0$ where $\beta(t)$ is given implicitly by

$$\int_{u(0)}^{\beta(t)} [f(s)]^{-1}\, ds = \delta t.$$

But $\int^\infty [f(s)]^{-1}\, ds = \infty$ implies that $\beta(t)$ exists for all $t \geq 0$. Thus, $u(x,t)$ exists for all $t \geq 0$, a contradiction. □

Condition (3.11) is satisfied by $f(u) = e^u$ and any positive $f(u)$ which grows at least as fast as $u^{1+\alpha}$, $\alpha > 0$, as $u \to \infty$. An important question is the following: Can one find a sufficient condition for blowup in finite time? Returning to IBVP (1.28)- (1.29), we can get a sufficient condition for a finite blowup time by the following comparison theorem.

Theorem 3.9 *Let $u(t)$ be the solution of the initial value problem*

$$u' = \delta e^u - \lambda_1 u,\ t \in (0,T) \text{ and } u(0) = 0$$

where λ_1 is the first eigenvalue of $-\Delta\phi = \lambda\phi$, $x \in \Omega$ and $\phi(x) = 0$, $x \in \partial\Omega$. Let $\theta(x,t)$ be the solution of IBVP (1.28)-(1.29) on $\overline{\Omega} \times [0,T)$; then

$$u(t) \leq \sup\{\theta(x,t) : x \in \overline{\Omega}\}$$

for $t \in [0,T)$.

Proof. Let ϕ be an eigenfunction associated with λ_1 where $\phi(x) \geq 0$ on Ω and $\int_\Omega \phi(x)\, dx = 1$. Define $a(t) = \int_\Omega \theta(x,t)\phi(x)\, dx$; then $a(t) \leq \sup\{\theta(x,t) : x \in \overline{\Omega}\}$. Multiplying (1.28) by $\phi(x)$ and integrating over Ω, we have

$$\begin{aligned} a'(t) &= \textstyle\int_\Omega \theta_t(x,t)\phi(x)\, dx \\ &= \textstyle\int_\Omega [\delta e^\theta \phi + \phi\Delta\theta]\, dx \\ &\geq \delta \exp(\textstyle\int_\Omega \theta\phi\, dx) + \textstyle\int_\Omega \theta\Delta\phi\, dx \\ &= \delta e^{a(t)} - \lambda_1 a(t) \end{aligned}$$

where Jensen's inequality and Green's identity have been used. Since $a'(t) \geq \delta e^{a(t)} - \lambda_1 a(t)$ and $a(0) = 0$, by a standard comparison theorem we have $u(t) \leq a(t) \leq \sup\{\theta(x,t) : x \in \overline{\Omega}\}$. □

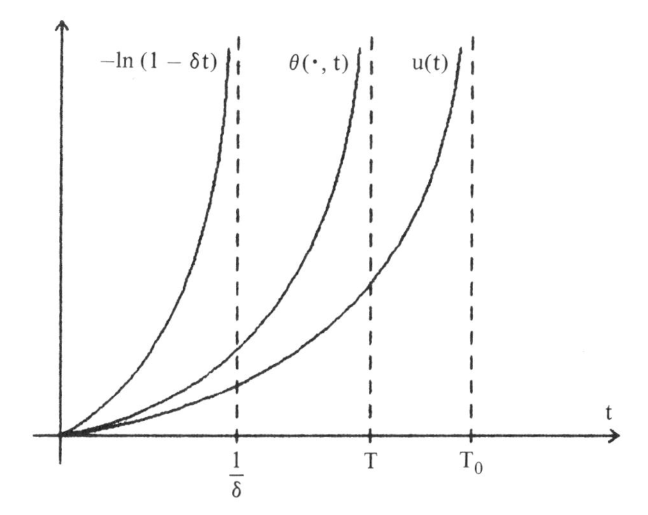

Figure 3.2.

Because of the elementary nature of the initial value problem in Theorem 3.9, we know that a unique solution $u(t)$ of these equations exists on $[0, T_0)$ with $u(t) \to \infty$ as $t \to T_0^-$ where

$$T_0 = \int_0^\infty \frac{dz}{\delta e^z - \lambda_1 z}.$$

Thus, $T_0 < \infty$ if $\delta > \delta^* := \lambda_1/e$ and we have the following implication.

Corollary 3.10 *If $\delta > \delta^* = \lambda_1/e$, then $T_0 < \infty$ and*

$$\lim_{t \to T^-} \sup\{\theta(x,t) : x \in \overline{\Omega}\} = \infty$$

where $T < T_0$. That is, blowup occurs in finite time.

The above corollary and inequalities are illustrated in Figure 3.2.
Table 3.1 gives the comparison between the critical value δ_{FK} and the value δ^*.

For a sphere $B_1 \subset \mathbb{R}^3$, the blowup time can be computed numerically by using the method of lines to solve IBVP (1.28)-(1.29) by approximating the spatial derivatives. The resulting system of first-order ordinary differential equations was integrated using a Runge-Kutta package RKF45. Table 3.2 uses $\Omega = B_1$, $\delta_{FK} = 3.32$, and $\delta^* = 3.63$:

Table 3.1.

Ω	δ_{FK}	λ_1	δ^*
S, slab	0.878	2.467	0.908
C, cylinder	2.000	5.784	2.128
B, sphere	3.320	9.872	3.631

Table 3.2.

δ	$\frac{1}{\delta}$	T_0	T
3.32	0.301	∞	
3.63	0.275	∞	
3.70	0.270	2.913	0.876
4.00	0.250	1.118	0.601
6.67	0.150	0.252	0.187
20.00	0.050	0.057	0.0503
50.00	0.020	0.021	0.0200

Of all solids of equal volume, the sphere has minimal surface area. An obvious conjecture is that for all solids Ω the sphere should explode first. The following comparison supports this conjecture. For the three solids of equal volume π: a sphere B, a parallelepiped P with edge length $\pi^{1/3}$, and a right circular cylinder C of radius 1 and height 1; the first eigenvalue λ_1 with $\Omega = B$,P, and C, respectively, is 11.656, 13.799, and 15.653. The values for δ^* are 4.288, 5.076, and 5.758, respectively. Numerical results are given in Table 3.3. This conjecture was proved by Bandle [BAN2] as

Theorem 3.11 *If IBVP (1.28)-(1.29) has a solution $\theta(x,t)$ on $\overline{\Omega} \times [0,T]$ where $\Omega = \{x : |x| < R\} = B_R$, then*

1. *IBVP (1.28)-(1.29) has a solution $u(x,t)$ on $\overline{\Omega} \times [0,T]$ for any other domain Ω of the same volume, and*
2. $\max\{u(x,t) : x \in \overline{\Omega}\} \leq \max\{\theta(x,t) : x \in \overline{B}_R\}$ *for all $t \in [0,T]$.*

For the more general initial-boundary value problem (3.7)-(3.8), we have the following result.

Theorem 3.12 *Consider IBVP (3.7)-(3.8) with $f(u) > 0$, $f'(u) \geq 0$, and $f''(u) \geq 0$ for $u \geq 0$, and $\int^\infty [f(u)]^{-1}\,du < \infty$. If $\delta > \delta^* := \lambda_1 \sup\{u/f(u) : u \geq 0\}$, then the unique solution $\theta(x,t)$ of (3.7)-(3.8) blows up in finite time T where*

$$\int_0^\infty [\delta f(z)]^{-1} dz < T < \int_0^\infty [\delta f(z) - \lambda_1 z]^{-1} dz =: T_0.$$

Table 3.3.

δ	$\frac{1}{\delta}$	T^B	T^P	T^C
4.288	0.233	∞	∞	∞
5.076	0.197	1.893	∞	∞
5.758	0.174	0.848	4.260	∞
20.000	0.050	0.095	0.098	0.101

Proof. The proof is the same as that for IBVP (1.28)-(1.29). Define $a(t) = \int_\Omega \phi(x)\theta(x,t)\,dx$ where ϕ is the nonnegative eigenfunction for λ_1 with $\int_\Omega \phi(x)\,dx = 1$. From (3.7) we have

$$a'(t) = \delta \int_\Omega \phi f(\theta)\,dx \;-\lambda_1 a(t)$$

for $t > 0$. Also, $a(0) = a_0 := \int_\Omega \phi u_0\,dx$.

By Jensen's inequality, $a' \geq \delta f(a) - \lambda_1 a$ for $t > 0$. If $\delta > \delta^* = \lambda_1 \sup\{a/f(a) : a \geq 0\}$, then $a(t) \to \infty$ as $t \to T_0^-$ where

$$T_0 = \int_0^\infty [\delta f(z) - \lambda_1 z]^{-1}\,dz < \infty.$$

Thus, $\sup\{\theta(x,t) : x \in \overline{\Omega}\} \to \infty$ as $t \to T^-$ with $T < T_0$. □

The last result shows that the blowup time is finite if $\delta > \delta^*$. What happens for $\delta \in (\delta_{FK}, \delta^*]$? The following is due to Lacey [LAC1].

Theorem 3.13 *If δ_{FK} is in the spectrum of (1.30)-(1.31) and if $\delta > \delta_{FK}$, then the unique solution $\theta(x,t)$ of IBVP (1.28)- (1.29) blows up in finite time T where*

$$T < \sqrt{\frac{2\pi^2}{\delta_{FK}(\delta - \delta_{FK})}}.$$

Proof. Let $w^*(x)$ be the solution of BVP (1.30)-(1.31) for $\delta = \delta_{FK}$. Then the first variational problem

$$\begin{aligned} -\Delta\phi &= \left[\delta_{FK}\, e^{w^*(x)}\right]\phi, \quad x \in \Omega \\ \phi(x) &= 0, \quad x \in \partial\Omega \end{aligned} \tag{3.12}$$

has a positive solution $\phi(x)$ on Ω such that $\int_\Omega \phi\,dx = 1$ (Amann, [AMA1]).

Define $v(x,t) = \theta(x,t) - w^*(x)$; then

$$\begin{aligned} v_t &= \theta_t \\ &= \delta e^\theta - \Delta\theta \\ &= (\delta - \delta_{FK})e^\theta + \delta_{FK} e^{w^*+v} + \Delta w^* + \Delta v \\ &= (\delta - \delta_{FK})e^\theta + \delta_{FK}(e^v - v - 1)e^{w^*} + \delta_{FK}\, v e^{w^*}. \end{aligned} \tag{3.13}$$

Set $a(t) = \int_\Omega \phi(x)v(x,t)\,dx$; then $a(t) \leq \sup\{\theta(x,t) : x \in \overline{\Omega}\}$ and $a(0) \geq -\sup\{w^*(x) : x \in \overline{\Omega}\}$. Multiply (3.13) by ϕ and integrate over Ω to obtain

$$\begin{aligned} a'(t) &= (\delta - \delta_{FK}) \int_\Omega \phi e^\theta dx + \delta_{FK} \int_\Omega \phi(e^v - v - 1)e^{w^*} dx \\ &\quad + \delta_{FK} \int_\Omega \phi v e^{w*} dx + \int_\Omega \phi\Delta v\,dx. \end{aligned}$$

By Green's identity and the fact that ϕ is a solution of (3.12), we have

$$\int_\Omega \phi \Delta v \, dx = \int_\Omega v \Delta \phi \, dx = -\delta_{FK} \int_\Omega v e^{w^*} \phi \, dx,$$

and thus,

$$a'(t) = (\delta - \delta_{FK}) \int_\Omega \phi e^\theta \, dx + \delta_{FK} \int_\Omega \phi (e^v - v - 1) e^{w^*} dx.$$

Since $(e^v - v - 1)e^{w^*} \geq \frac{1}{2} v^2$, and by Jensen's inequality, we have that

$$\delta_{FK} \int_\Omega \phi (e^v - v - 1) e^{w^*} dx \geq \frac{1}{2} \delta_{FK} \int_\Omega \phi v^2 \, dx \geq \frac{1}{2} \delta_{FK} a^2$$

and clearly $\int_\Omega \phi e^\theta \, dx \geq 1$. Thus, $a(t)$ satisfies the differential inequality

$$a'(t) \geq (\delta - \delta_{FK}) + \frac{1}{2} \delta_{FK} a^2(t)$$

with

$$a(0) \geq -\sup\{w^*(x) : x \in \overline{\Omega}\} =: -w_m^*.$$

The solution of $z' = (\delta - \delta_{FK}) + \frac{1}{2}\delta_{FK} z^2$, $z(0) = -w_m^*$ is

$$z(t) = \frac{2}{c\delta_{FK}} \tan\left(\frac{t}{c} - \tan^{-1}(cw_m^*)\right)$$

where

$$c = \sqrt{\frac{2}{\delta_{FK}(\delta - \delta_{FK})}}.$$

The function $z(t)$ blows up before $t_A = c\pi$. Thus, $\sup\{\theta(x,t) : x \in \overline{\Omega}\} \geq a(t) \geq z(t)$ and $T \leq t_A$. □

This result has been extended by Lacey [LAC1] to IVBP (3.7)-(3.8) as follows.

Theorem 3.14 *Consider IBVP (3.7)-(3.8) with $f(u) > 0$, $f'(u) \geq 0$, and $f''(u) \geq 0$ for $u \geq 0$. Assume that $\int^\infty [f(s)]^{-1} \, ds < \infty$. If δ_{FK} is in the spectrum of (3.9)-(3.10), then the unique solution $u(x,t)$ of IBVP (3.7)-(3.8) for $\delta > \delta_{FK}$ blows up in finite time (in the L^∞-sense) at*

$$T < A + B(\delta - \delta_{FK})^{-1/2}$$

where the constants A and B are independent of δ.

Proof. Let w be the solution of:

$$-\Delta w = \delta_{FK} f(w), \quad x \in \Omega$$
$$\frac{\partial w}{\partial \eta} + \beta w = 0, \quad x \in \partial\Omega.$$

Let ϕ be the solution of the first variational problem

$$-\Delta \phi = \delta_{FK} f'(w)\phi, \quad x \in \Omega$$
$$\frac{\partial \phi}{\partial \eta} + \beta \phi = 0, \quad x \in \partial\Omega.$$

Set $v = u - w$ where u is a solution of (3.7)-(3.8). Define the function $a(t) = \int_\Omega \phi(x)v(x,t)\,dx$; then

$$a'(t) \geq (\delta - \delta_{FK})I + \delta_{FK}\int_\Omega \phi(x)[f(w+v) - f(w) - vf'(w)]\,dx \quad (3.14)$$

where $I = f(u_B)$ and $u_B = \min\{0, \inf\{u_0(x) : x \in \overline{\Omega}\}\}$.

Let $g(s) = K_1[f(s) - f(0) - sf'(0)]$ for $K_1 > 0$ sufficiently small; then $g(0) = g'(0) = 0$, $g(s)$ is convex, $g(s) \leq f(w+s) - f(w) - sf'(w)$ for $x \in \overline{\Omega}$ and for $\min\{v(x) : x \in \overline{\Omega}\} \leq s \leq \max\{v(x) : x \in \overline{\Omega}\}$, and $\int^\infty \frac{ds}{f(s)} < \infty$.

Equation (3.14) implies

$$a'(t) \geq (\delta - \delta_{FK})I + \delta_{FK}\int_\Omega \phi g(v)\,dx \geq (\delta - \delta_{FK})I + \delta_{FK}\,g(a)$$

where the last inequality follows from Jensen's inequality ($\int_\Omega \phi g(v)\,dx \geq g(\int_\Omega \phi v\,dx)$). For

$$a_0 = \int_\Omega \phi(u_0 - w)\,dx \text{ and } a_1 < \min\{0, a_0\},$$

choose a_2 such that $0 < a_2 < -a_1$. Let $h(s) = [(\delta - \delta_{FK})I + \delta_{FK}g(s)]^{-1}$; then

$$\begin{aligned} t &\leq K_2 + \int_{a_1}^{a(t)} h(s)\,ds \\ &< \int_{a_1}^{-a_2} h(s)\,ds + \int_{-a_2}^{a_2} h(s)\,ds + \int_{a_2}^{\infty} h(s)\,ds + K_2. \end{aligned} \quad (3.15)$$

The sum of the first and third integrals on the right-hand side of (3.15) is bounded by the quantity

$$K_3 = \delta_{FK}\left(\int_{a_1}^{-a_2} [g(s)]^{-1}\,ds + \int_{a_2}^{\infty} [g(s)]^{-1}\,ds\right).$$

The second integral is bounded by

$$\int_{-\infty}^{\infty} [(\delta - \delta_{FK})I + K_4 s^2]^{-1}\,ds = [(\delta - \delta_{FK})IK_4]^{-1/2}\pi$$

where $K_4 = \frac{1}{2}\delta_{FK} \inf\{g''(s) : |s| < a_2\}$.

We can deduce that $u(x,t)$ blows up at $T < A + B(\delta - \delta_{FK})^{-1/2}$ where $A = K_2 + K_4$ and $B = \pi(IK_4)^{-1/2}$. □

The requirement that the critical value δ_{FK} be in the spectrum of problem (3.7)-(3.8) may not be necessary. The following result due to Bellout [BEL] does not make such an assumption. However, a concavity assumption replaces the spectral condition.

Theorem 3.15 *Consider IBVP (3.7)-(3.8) where the mixed boundary condition is modified to*

$$\alpha \frac{\partial u}{\partial \eta} + \beta u = 0, \quad (x,t) \in \partial\Omega \times (0,\infty)$$

where α and β are nonnegative constants such that $\alpha + \beta > 0$. Suppose that the function $f \in C^3([0,\infty))$ satisfies the conditions

$$f(0) > 0, \; f'(s) > 0 \;\; for \;\; s > 0, \quad \int^{\infty} \frac{ds}{f(s)}\, ds < \infty, \;\; and \;\; \left(\frac{f}{f'}\right)'' \leq 0.$$

In addition, if $\alpha\beta \neq 0$, assume that

$$\frac{d}{ds}\left[\frac{sf'(s)}{f(s)}\right] \geq 0.$$

If $\delta' := \delta - \delta_{FK} > 0$, then the solution to (3.7)- (3.8) blows up in finite time $T \leq \frac{K}{\delta'}$ where K is a positive constant dependent on δ, δ_{FK}, and $M := \int_0^\infty [f(s)]^{-1}\, ds$.

Proof. Consider the problem

$$v_t - \Delta v = \delta a^2 t^2 f(v), \; (x,t) \in \Omega \times (0, T_1) \tag{3.16}$$

with initial-boundary conditions

$$\begin{aligned} &v(x,0) = 0, \quad x \in \Omega \\ &\alpha \tfrac{\partial v}{\partial \eta} + \beta v = 0, \quad (x,t) \in \partial\Omega \times (0, T_1) \end{aligned} \tag{3.17}$$

where

$$T_1^{-1} = \frac{\delta'}{M} \frac{4\delta_{FK} + \delta'}{4(2\delta_{FK} + \delta')}\left[1 - \left(\frac{4\delta_{FK} + \delta'}{2(2\delta_{FK} + \delta')}\right)^{1/2}\right] =: \frac{\delta'}{K} =: a.$$

There is a unique solution $v(x,t)$ to this problem such that v ceases to exist only by becoming infinite. We wish to prove that v blows up in a finite time $T_0 \leq T_1$. Without loss of generality, assume that v is finite in $\Omega \times [0, T_0)$. Also assume that $v \in C^{6,3}(\overline{\Omega} \times [0, T_0))$ and $v_t/f(v) \in C^{2,1}(\overline{\Omega} \times [0, T_0))$.

Define $w(x,t) = \int_0^v [f(s)]^{-1}\, ds$ and $z(x,t) = w_{tt}(x,t)$. We wish to show that $z(x,t) \geq 0$. From equation (3.16) we see that w satisfies the equation

$$w_t = \Delta w + |\nabla w|^2 f'(v) + \delta a^2 l^2. \tag{3.18}$$

Define

$$L(z) = z_t - \Delta z - 2f'\nabla z \bullet \nabla w - f'' f' z \nabla w \bullet \nabla w.$$

Differentiating equation (3.18) twice with respect to t yields

$$L(z) = 2\delta a^2 + 2f'|\nabla w_t|^2 + (w_t)^2(f''' f^2 + f'' f' f) + 4w_t f'' f \nabla w \bullet \nabla w_t.$$

Using

$$|w_t \nabla w \bullet \nabla w_t| \leq \frac{1}{2}\left[|\nabla w|^2 w_t^2 |\varepsilon| + \frac{1}{|\varepsilon|}|\nabla w_t|^2\right]$$

we obtain for $\varepsilon = f'' f/f'$ the inequality

$$L(z) \geq 2\delta a^2 - f(f')^2 (\frac{f}{f'})''(w_t)^2 |\nabla w|^2.$$

Applying the assumption that f/f' is concave, we have that $L(z) \geq 2\delta a^2 > 0$.

From equation (3.17) we have that $z(x,0) = 0$. If $\alpha = 0$ (respectively $\beta = 0$), then $z(x,t) = 0$ (respectively $\frac{\partial z}{\partial \eta} = 0$) on $\partial\Omega \times (0, T_0)$. Since the coefficients of L remain bounded as long as v is bounded, the maximum principle implies that $z(x,t) \geq 0$ for $t \in [0, T_0)$.

In the case $\alpha\beta \neq 0$, let $b = \beta/\alpha$; then $\frac{\partial v}{\partial \eta} = -bv$ on $\partial\Omega \times (0, T_0)$. Differentiating with respect to t and dividing by f we obtain

$$\frac{v_{\eta t}}{f} = -b\frac{v_t}{f}. \tag{3.19}$$

Also,

$$\frac{\partial}{\partial \eta}\frac{v_t}{f} = \frac{v_{\eta t}}{f} - \frac{v_t v_\eta f'}{f^2} = \frac{v_{\eta t}}{f} + b\frac{v v_t f'}{f^2}.$$

Substituting into equation (3.19) yields

$$\frac{\partial}{\partial \eta}\frac{v_t}{f} = b\left(\frac{v_t}{f}\right)\left[v\left(\frac{f'}{f}\right) - 1\right].$$

Since $w_t = v_t/f$, we obtain

$$w_{\eta t} = bw_t \left[v\left(\frac{f'}{f}\right) - 1\right].$$

Differentiating with respect to t and using $z = w_{tt}$ we obtain

$$z_{eta} = bz[v(f'/f) - 1] + b(w_t)^2 v f'[\ln(vf'/f)]' \geq bz[v(f'/f) - 1] \tag{3.20}$$

where the condition $[sf'(s)/f(s)]' \geq 0$ was used.

Since $v(x,0) = 0$, there exists an $\varepsilon > 0$ such that the right-hand side of the inequality in (3.20) has a negative coefficient for z on the set $\overline{\Omega} \times (0,\varepsilon)$. We also had $L(z) > 0$ on $\overline{\Omega} \times (0,\varepsilon)$ and $z(x,0) = 0$ on Ω. Combining this with equation (3.20), the strong maximum principle implies that $z(x,t) > 0$ on $\overline{\Omega} \times (0,\varepsilon)$.

If z becomes negative somewhere in $\Omega \times (0,T_0)$, then let t^* be the smallest positive time where z becomes zero. Since $L(z) > 0$, it must be that $z(x,t^*) > 0$ on Ω. This forces z to have a zero at the point $x^* \in \partial\Omega$. By the maximum principle, $z_\eta(x^*,t^*) < 0$; but equation (3.20) implies that $z_\eta(x^*,t^*) \geq 0$, a contradiction. Thus, $z(x,t) \geq 0$ on $\Omega \times (0,T_0)$.

Using this information we can show that $v(x_0,t) \to \infty$ as $t \to T_0$ for some $x_0 \in \Omega$ and for some $T_0 \in (0,T_1]$. Define

$$t_0 = \frac{1}{a}\left(\frac{\delta_{FK} + \frac{1}{4}\delta'}{\delta_{FK} + \frac{1}{2}\delta'}\right)^{1/2};$$

then $a^2 t_0^2(\delta_{FK} + \frac{1}{2}\delta') = \delta_{FK} + \frac{1}{4}\delta'$. At $t = t_0$ equation (3.16) can be written as

$$v_t = \Delta v + (\delta_{FK} + \frac{1}{4}\delta')f(v) + \frac{1}{2}\delta' a^2 t_0^2 f(v). \tag{3.21}$$

The steady-state problem (3.9)-(3.10) has no solution for $\delta' > 0$, in particular, no solution for $\delta = \delta_{FK} + \frac{1}{4}\delta'$. In fact, the steady-state problem has no upper solution (Amann, [AMA1]), so there exists an x_0 such that at (x_0,t_0), $\Delta v + (\delta_{FK} + \frac{1}{4}\delta')f(v) > 0$. Consequently, using the result $z \geq 0$ in (3.21) we have

$$\frac{v_t}{f(v)} \geq \frac{1}{2}\delta' a^2 t_0^2$$

at (x_0,t) for $t \geq t_0$. Integrating with respect to t yields

$$\int_0^{v(x_0,T_0)} [f(s)]^{-1}\, ds \geq \frac{1}{2}\delta' a^2 t_0^2 (T_0 - t_0). \tag{3.22}$$

Either v becomes infinite at $T_0 < T_1$, or, $v(x_0,t)$ is finite for all $t \in [0,T_1)$. In the latter case, the right-hand side of the inequality in (3.22) as $T_0 \to T_1$ becomes the value $M = \int_0^\infty [f(s)]^{-1}\, ds$. This forces $v(x_0,T_1) = \infty$.

Finally, to prove that the solution $u(x,t)$ to (3.7)-(3.8) becomes infinite in finite time, consider the following. Since $a^2t^2 \leq 1$ for $t \leq T_0$, v satisfies

$$v_t \leq \Delta v + \delta f(v), \quad (x,t) \in \Omega \times (0,T_0).$$

The function $w = u - v$ satisfies:

$$w_t \geq \Delta w + \gamma w, \quad (x,t) \in \Omega \times (0,T_0)$$

$$\alpha w_\eta + \beta w = 0, \quad (x,t) \in \partial\Omega \times (0,T_0),$$

where $\gamma = f'(\theta u + (1-\theta)v)$ is a bounded function as long as u and v are both bounded. By the maximum principle, we see that $w \geq 0$. Thus, $u \geq v$, and u must become infinite at some time $T \leq T_0$. □

Theorem 3.15 shows that blowup in finite time occurs for IBVP (3.7)-(3.8) for arbitrary Ω for any $\delta > \delta_{FK}$. An open question is: Can the concavity assumption on f/f' be dropped?

Blowup can occur in finite time even for $\delta < \delta_{FK}$ if the initial data function $u_0(x) \geq 0$ is sufficiently large [LAC1]. To illustrate this, assume $f(u) = e^u$ in IBVP (1.28)-(1.29). Assume $w(x)$ is a nonminimal solution of (1.30)-(1.31), the associated steady-state model. (For $\delta < \delta_{FK}$ sufficiently close to δ_{FK}, this is often possible [DEF].) Let w_m be the minimal solution. By a result of Amann [AMA2], the principle eigenvalue λ_1 of

$$\Delta\phi + (\lambda + \delta e^w)\phi = 0, \quad x \in \Omega$$

$$\phi(x) = 0, \quad x \in \partial\Omega$$

is nonpositive. Let $\xi(x)$ be the associated eigenfunction with $\xi(x) > 0$ and $\int_\Omega \xi(x)\,dx = 1$. Set $v(x,t) = u(x,t) - w(x)$ and $a(t) = \int_\Omega v\xi\,dx$; then

$$a(0) = \int_\Omega [u_0(x) - w(x)]\xi(x)\,dx \text{ and } a(t) \leq \sup_\Omega u(x,t),$$

with

$$\begin{aligned} v_t &= \Delta u + \delta e^{w+v} - \Delta w - \delta e^w \\ &= \Delta v + \delta[e^{w+v} - e^w - ve^w] + \delta e^w v - \lambda_1 v + \lambda_1 v. \end{aligned}$$

This implies

$$\begin{aligned} a'(t) &= \textstyle\int_\Omega \xi \Delta v\,dx + \delta \int_\Omega [e^{w+v} - e^w - ve^w]\xi\,dx \\ &\quad + \textstyle\int_\Omega (\lambda_1 + \delta e^w) v\xi\,dx - \lambda_1 \int_\Omega v\xi\,dx \\ &= -\lambda_1 a + \delta \textstyle\int_\Omega e^w \xi [e^v - v - 1]\,dx \\ &\geq -\lambda_1 a + \frac{\delta}{2} \textstyle\int_\Omega e^w \xi v^2\,dx. \end{aligned}$$

By Jensen's inequality, $a'(t) \geq -\lambda_1 a + Ka^2$ where $K > 0$. If $a(0) > 0$, then $a(t)$ satisfies $a(t) \geq \alpha(t)$ where $\alpha(t)$ is the solution of

$$\alpha' = K\alpha^2 - \lambda_1\alpha, \ t > 0, \ \ \alpha(0) = a(0).$$

Since $-\lambda_1 \geq 0$, $\int^\infty \frac{dz}{Kz^2 - \lambda_1 z} < \infty$ and $\alpha(t)$ blows up in finite time. Hence, the solution $u(x,t)$ of (1.28)-(1.29) blows up in finite time provided

$$\int_\Omega \xi(x) u_0(x)\,dx > \int_\Omega \xi(x) w(x)\,dx.$$

3.3 Blowup: Where?

In this section we deal with the topic of where blowup occurs for solutions to the solid fuel model. We consider the partial differential equation

$$u_t - \Delta u = f(u), \quad (x,t) \in \Omega \times (0,T) \tag{3.23}$$

with initial-boundary conditions

$$\begin{aligned} u(x,0) &= \phi(x), \quad x \in \Omega \\ u(x,t) &= 0, \quad (x,t) \in \partial\Omega \times (0,T) \end{aligned} \tag{3.24}$$

where $\Omega = \{x \in \mathbb{R}^n : |x| < R\}$.

We assume that $\phi \in C^2(\overline{\Omega})$ is a radially symmetric function, say $\phi = \phi(r)$ where $r = |x|$. In addition, we assume that $\phi'(r) \leq 0$ for $r \in [0, R]$ and $\phi(R) = 0$. Consequently, $\phi'(0) = 0$, $\phi''(0) \leq 0$, and $\phi(r) \geq 0$.

We also assume that $f \in C^2(\mathbb{R})$, $f(u) > 0$ for $u > 0$, $f'(u) \geq 0$, $f''(u) \geq 0$, and $\int^\infty \frac{du}{f} < \infty$. For example, the functions $\exp(u)$ and $(u+\lambda)^p$, $(\lambda \geq 0,\ p > 1)$, satisfy these conditions.

By uniqueness and since ϕ is radially symmetric, for each $t \geq 0$ the solution $u(\cdot, t)$ of (3.23)-(3.24) is radially symmetric. By the maximum principle and since ϕ is radially decreasing, the solution $u(\cdot, t)$ is radially decreasing. Therefore, we consider the equivalent formulation

$$u_t - \left(u_{rr} + \frac{n-1}{r} u_r\right) = f(u), \quad (r,t) \in (0,R) \times (0,T)$$

with initial-boundary conditions

$$\begin{aligned} &u(r,0) = \phi(r), \quad r \in (0,R) \\ &u(R,t) = 0, \ u_r(0,t) = 0, \quad t \in (0,T) \end{aligned}$$

A unique solution of (3.23)-(3.24) exists for $t \in [0, \sigma)$ for $\sigma > 0$ sufficiently small. By the maximum principle, $U(t) := \sup_{r \in (0,R)} u(r,t)$ is an increasing function. Define the value

$$T = \sup\{\sigma > 0 : \text{(3.23)-(3.24) has a solution } u(\cdot, t) \text{ for } t \in [0, \sigma)\}.$$

If $T < \infty$, then $U(T^-) = \infty$. Otherwise, if $U(T^-) < \infty$, then the solution to (3.23)-(3.24) could be extended to a time interval $[0, \sigma + \varepsilon)$ with $\varepsilon > 0$ by using standard parabolic estimates, a contradiction to the maximality of T.

We will assume that the necessary conditions are met for (3.23)-(3.24) to have a blowup time $T < \infty$.

Definition 3.3 *A point $x \in \Omega$ is a* blowup point *for (3.23)-(3.24) if there exists a sequence $\{(x_m, t_m)\}_{m=0}^{\infty}$ such that $t_m \to T^-$, $x_m \to x$, and $u(x_m, t_m) \to \infty$ as $m \to \infty$.*

The first theorem we prove shows that if the nonlinearity $f(u)$ satisfies a certain condition, then blowup occurs at a single point ($x = 0$).

Theorem 3.16 *Suppose there is a function $F(u)$ such that $F \geq 0$, $F' \geq 0$, $F'' \geq 0$, $\int^{\infty} \frac{ds}{F(s)} < \infty$, and*

$$f'F - F'f \geq 2\varepsilon FF' \tag{3.25}$$

for $\varepsilon > 0$ sufficiently small; then the only blowup point for (3.23)-(3.24) is the point $x = 0$.

Proof. We wish to get a lower bound on $u_r(r,t)$. We already know that $u_r(r,t) \leq 0$ for $r \in (0, R)$.

Define the function $J(r,t) = r^{n-1}u_r + \varepsilon r^n F(u)$. We will show that $J(r,t) \leq 0$ for $(r,t) \in \Omega \times (0,T)$. It can be shown that J is a solution to

$$\begin{aligned} &J_t + \tfrac{n-1}{r}J_r - J_{rr} + cJ \\ &\quad = \varepsilon r^{n-1}(F'f - Ff' + 2\varepsilon FF') - \varepsilon^2 r^{n+2}F^2F'' - \tfrac{\varepsilon F''}{r^{n-1}}J^2 \end{aligned}$$

where $c = 2\varepsilon F' - 2\varepsilon^2 FF'' - f'$. As long as F satisfies the condition in (3.25), we have

$$J_t + \frac{n-1}{r}J_r - J_{rr} + cJ \leq 0.$$

Note that $J(0,t) = 0$ for $t \geq 0$. If $\phi'(r) < 0$ for $r \in (0, R]$, and $\phi''(0) < 0$, then for $\varepsilon > 0$ sufficiently small,

$$J(r,0) = r^{n-1}\phi'(r) + \varepsilon r^n F(\phi(r)) < 0 \tag{3.26}$$

for $r \in (0, R]$. [If $\phi'(r) = 0$ for some $r > 0$ or if $\phi''(0) = 0$, then equation (3.26) is no longer valid. See the note at the end of the theorem for the necessary modification to include these cases.] Finally, for $\varepsilon > 0$ sufficiently small,

$$\begin{aligned} J_r(R,t) &= R^{n-1}[u_t(R,t) - f(0)] + \varepsilon R^{n-1}u_r(R,t)F'(0) + \varepsilon n R^{n-1}F(0) \\ &\leq R^{n-1}[\varepsilon n F(0) - f(0)] \leq 0 \end{aligned}$$

where we have used $u_t(R,t) = 0$ and $u_r(R,t) \leq 0$. By the maximum principle, $J(r,t) \leq 0$ for $(r,t) \in (0,R) \times (0,T)$. As a result we have the inequality

$$u_r(r,t) \leq -\varepsilon r F(u(r,t)) \tag{3.27}$$

for $(r,t) \in (0,R) \times (0,T)$.

Define the function

$$G(u) = \int_u^\infty \frac{1}{F(s)}\, ds$$

for $u > 0$. The condition (3.27) yields $[G(u)]_r \geq \varepsilon r$. An integration yields

$$G(u(r,t)) \geq G(u(0,t)) + \frac{\varepsilon}{2} r^2 \geq \frac{\varepsilon}{2} r^2. \tag{3.28}$$

If there is a blowup point at some $r > 0$, then $u(r,t) \to \infty$ as $t \to T^-$ so that $G(u(r,t)) \to 0$ as $t \to T^-$. This is a contradiction to (3.28), so the only blowup point is at $r = 0$.

In the event that $\phi'(r) = 0$ for some $r > 0$ or $\phi''(0) = 0$, we can make the following modification to the proof. By the maximum principle, $u_{x_1}(x,t) < 0$ on the set $[\Omega \cap \{x : x > x_1\}] \times (0,T)$. Also, $u_{x_1}(0,t) = 0$ and $u_{x_1 x_1}(0,t) < 0$. Define $\Phi_\eta(r) = u(r,\eta)$ for any $\eta \in (0,T)$; then $\Phi'_\eta(r) < 0$ for $r \in (0,R]$ and $\Phi''_\eta(r) < 0$. Consequently, equation (3.26) can be replaced by

$$J(r,\eta) = r^{n-1}\Phi'_\eta(r) + \varepsilon r^n F(\Phi_\eta(r)) < 0.$$

The remainder of the proof of Theorem 3.16 is the same except that we conclude by the maximum principle that $J(r,t) \leq 0$ on $(0,R) \times (\eta, T)$ for all $\eta \in (0,T)$. Thus, $J(r,t) \leq 0$ on the entire set $(0,R) \times (0,T)$. □

Corollary 3.17 *Under the assumptions of Theorem 3.16, (3.23)-(3.24) has only the blowup point $r = 0$ for the special cases $f(u) = e^u$ and $f(u) = (u+\lambda)^p$, ($\lambda \geq 0$, $p > 1$). Moreover:*

If $f(u) = e^u$, then for any $\alpha \in (0,1)$,

$$u(r,t) \leq -\frac{1}{\alpha} \ln\left(\frac{\alpha \varepsilon r^2}{2}\right) \tag{3.29}$$

or

$$u(r,t) \leq -2\ln(r) + \ln(\ln r^{-1}) + C \tag{3.30}$$

for $(r,t) \in (0,R) \times (0,T)$ and for some constant C.

If $f(u) = (u+\lambda)^p$ for $p > 1$ and $\lambda \geq 0$, then for any $\gamma \in (1,p)$,

$$u(r,t) \leq \left(\frac{(\gamma-1)\varepsilon r^2}{2}\right)^{\frac{1}{1-\gamma}} \tag{3.31}$$

for $(r,t) \in (0,R) \times (0,T)$.

Proof. For $f(u) = e^u$, choose $F(u) = e^{\alpha u}$. The condition (3.25) appears as:

$$(1-\alpha)e^{(1+\alpha)u} \geq 2\alpha\varepsilon e^{2\alpha u}$$

which is valid for $\alpha \in (0,1)$ and for $\varepsilon \leq (1-\alpha)/2\alpha$. The condition that $\varepsilon n F(0) \leq f(0)$ requires that $\varepsilon < 1/n$. From (3.28) we conclude that

$$\frac{1}{\alpha} e^{-\alpha u} \geq \frac{\varepsilon}{2} r^2$$

from which the bound (3.29) on u follows. One could also choose $F(u) = \frac{e^u}{u+1}$. Condition (3.25) will be valid as long as $\varepsilon \leq 1/2$ and $\varepsilon n F(0) \leq f(0)$ requires $\varepsilon < 1/n$. Inequality (3.28) becomes

$$(u+2)e^{-u} \geq \frac{\varepsilon}{2} r^2.$$

For $r > 0$ sufficiently small and for $t < T$ sufficiently close to T, we have $u + 2 \leq 2u$, so

$$2ue^{-u} \geq (u+2)e^{-u} \geq \frac{\varepsilon}{2} r^2.$$

Taking the logarithm and using the previous estimate and (3.29) on $u(r,t)$ gives us the estimate

$$\begin{aligned} u(r,t) &\leq \ln(u) - 2\ln(r) + C_1 \\ &\leq \ln\left(\tfrac{2}{\alpha}\ln(\tfrac{1}{r}) + K_\alpha\right) - 2\ln(r) + C_1 \\ &\leq -2\ln(r) + \ln(\ln r^{-1}) + C \end{aligned}$$

for some constant C and where r is sufficiently small and t is sufficiently close to T.

For $f(u) = (u+\lambda)^p$ where $p > 1$ and $\lambda > 0$, choose $F(u) = (u+\lambda)^\gamma$. The condition (3.25) appears as:

$$(p-\gamma)(u+\lambda)^{p+\gamma-1} \geq 2\varepsilon\gamma(u+\lambda)^{2\gamma-1}$$

which is valid for $\gamma \in (0,p)$ and for $\varepsilon \leq (p-\gamma)\lambda^{p-\gamma}/2\gamma$. The condition that $\varepsilon n F(0) \leq f(0)$ requires that $\varepsilon < \lambda^{p-\gamma}/n$. From (3.28) we conclude that

$$\frac{1}{p-1}(u+\lambda)^{1-p} \geq \frac{\varepsilon}{2} r^2$$

from which the bound (3.31) on u follows. For $\lambda = 0$, the proof of Theorem 3.16 can be modified by choosing $J = r^{n-1}u_r + \varepsilon r^{n+\delta}F(u)$ for $\delta > 0$ and small. The bound on $u(r,t)$ for this case is constructed just as in the case $\lambda > 0$. □

The next results are on "in-time" growth rates. We need not assume that Ω is a ball nor that $\phi(x)$ is radially symmetric for these results.

Theorem 3.18 *For any bounded domain Ω, the function*

$$U(t) = \sup\{u(x,t) : x \in \Omega\}$$

is Lipschitz continuous and $U'(t) \leq f(U(t))$ a.e.

Proof. Let $t, t_0 \in [0,T)$. There are points $\overline{x}, x_0 \in \Omega$ such that $U(t) = u(\overline{x}, t)$ and $U(t_0) = u(x_0, t_0)$. It follows that

$$U(t) - U(t_0) \geq u(x_0,t) - u(x_0,t_0) = (t-t_0)u_t(x_0,t_0) + \mathbf{o}(t-t_0)$$

and

$$U(t) - U(t_0) \leq u(\overline{x}, t) - u(\overline{x}, t_0) = (t - t_0)u_t(\overline{x}, t) + \mathbf{o}(t - t_0)$$

which imply that $U(t)$ is Lipschitz continuous and hence differentiable almost everywhere.

For $t > t_0$, we have

$$\frac{U(t) - U(t_0)}{t - t_0} \leq u_t(\overline{x}, t) + \mathbf{o}(1) = \Delta u(\overline{x}, t) + f(u(\overline{x}, t)) + \mathbf{o}(1).$$

But at $(\overline{x}, t)$ we have a maximum for u, so $\Delta u(\overline{x}, t) \leq 0$. Letting $t_0 \to t$ gives us $U'(t) \leq f(U(t))$ a.e. □

Corollary 3.19 *Let $u(x,t)$ be a solution to (3.23)-(3.24). If $f(u) = e^u$, then*

$$U(t) \geq -\ln(T - t) \quad \textit{for } t \in (0, T).$$

If $f(u) = (u + \lambda)^p$ for $p > 1$ and $\lambda \geq 0$, then

$$U(t) + \lambda \geq \left(\frac{\beta}{T - t}\right)^{\beta} \quad \textit{for } t \in (0, T)$$

where $\beta = \frac{1}{p-1}$.

Proof. Integration of $U'(t) \leq f(U(t))$ yields

$$\int_{U(t)}^{\infty} \frac{1}{f(s)}\, ds \leq T - t.$$

The inequalities easily follow by an integration. □

We now have a lower bound on $U(t)$ for solutions to (3.23)-(3.24). To obtain an upper bound for solutions $u(x,t)$ requires a few more assumptions.

Theorem 3.20 *Let Ω be any bounded domain. Let the initial data $\phi(x)$ satisfy $\Delta\phi + f(\phi) \geq 0$. Assume that the set of blowup points of (3.23)-(3.24) is compact. For any $\eta > 0$ there exists a $\xi > 0$ such that*

$$u_t \geq \xi f(u) \quad \textit{for } (x,t) \in \Omega^{\eta} \times (\eta, T)$$

where $\Omega^{\eta} = \{x \in \Omega : \mathrm{dist}(x, \partial\Omega) > \eta\}$.

Proof. Consider the function $J(x,t) = u_t - \xi f(u)$ on $\Omega^{\eta} \times (\eta, T)$ where $\xi > 0$ is to be determined. We know that $u_t > 0$ on $\Omega^{\eta} \times (\eta, T)$ by the maximum principle. Also, J satisfies

$$J_t - \Delta J - f'(u)J = \xi f''(u)|\nabla u|^2 \geq 0.$$

Since the set of blowup points is compact, if η is sufficiently small,

$$f(u) \leq C_1 < \infty \text{ for } (x,t) \in \partial\Omega^\eta \times (0,T).$$

Also, $u_t \geq C_2 > 0$ on the parabolic boundary of $\Omega^\eta \times (\eta, T)$. For $\xi > 0$ sufficiently small, $J \geq C_2 - \xi C_1 > 0$ on the parabolic boundary. By the maximum principle,

$$J \geq 0 \text{ on } \Omega^\eta \times (\eta, T)$$

and so $u_t \geq \xi f(u)$ on this set. □

Corollary 3.21 *Let $u(x,t)$ be a solution to (3.23)-(3.24). If $f(u) = e^u$, then*

$$u(x,t) \leq -\ln(T-t) - \ln\xi \text{ for } (x,t) \in \Omega \times (0,T).$$

for $\xi > 0$ sufficiently small. If $f(u) = (u+\lambda)^p$ for $p > 1$ and $\lambda \geq 0$, then

$$u(x,t) + \lambda \leq \left(\frac{\beta}{\xi(T-t)}\right)^\beta \text{ for } (x,t) \in \Omega \times (0,T)$$

for $\xi > 0$ sufficiently small where $\beta = \frac{1}{p-1}$.

3.4 Blowup: How?

We again work with the initial-boundary value problem (3.23)-(3.24) on a ball $\Omega = \{\xi \in \mathbb{R}^n : |\xi| < R\}$:

$$u_t - \left(u_{rr} + \tfrac{n-1}{r}u_r\right) = f(u), \quad (r,t) \in (0,R) \times (0,T)$$
$$u(r,0) = \phi(r), \quad r \in (0,R)$$
$$u(R,t) = 0, \quad u_r(0,t) = 0, \quad t \in (0,T).$$

The assumptions on the initial data $\phi(r)$ remain the same. That is, $\phi \in C^2([0,R])$, $\phi'(r) \leq 0$ for $r \in [0,R]$, and $\phi(R) = 0$. In addition, assume that $\Delta\phi + f(\phi) \geq 0$ for $r \in (0,R)$. This will guarantee that $u_t(r,t) \geq 0$ (see Theorem 3.3). The function $f(u)$ will be either e^u or u^p for $p > 1$.

We assume that (3.23)-(3.24) has finite blowup time T. In Section 3.3 we proved that blowup can only occur at $r = 0$. We want to analyze the asymptotic behavior of $u(r,t)$ near $r = 0$ as $t \to T^-$.

For $n = 1$ and $f(u) = e^u$, Kassoy and Poland [KAS3],[KAS4] and Kapila [KAP1] argued the following formal final time analysis. Since $u(r,t)$ blows up only at $r = 0$, $u(r,t) \to U_F(r)$ for $r \in (0,R]$ and U_F should be describable.

Let $\tau = T - t$, $x = r\tau^{-1/2}$, and $\theta(x,\tau) = u(r,t)$. A formal power series expansion in τ,

$$\theta(x,\tau) = -\ln\tau + y(x) + \sum_{k=1}^{\infty} \tau^k y_k(x), \tag{3.32}$$

was postulated. As $\tau \to 0^+$ we have the asymptotic condition

$$\theta(x,\tau) \sim -\ln\tau + y(x).$$

We wish to describe the function $y(x)$.

Substituting (3.32) into (3.23) and letting $\tau \to 0^+$ yields the differential equation

$$y'' + \left(\frac{n-1}{x} - \frac{x}{2}\right)y' + e^y - 1 = 0, \quad x > 0.$$

By Corollary 3.19, $u(0,t) = \max_{r\in[0,R]} u(r,t) = U(t) \geq -\ln(T-t)$ and as a result,

$$0 \leq u(0,t) + \ln(T-t) = \theta(0,\tau) + \ln\tau = y(0) + \sum_{k=1}^{\infty} \tau^k y_k(0)$$

so that as $\tau \to 0^+$ we have $y(0) \geq 0$. Also, we have $u_r(r,t) = \tau^{-1/2}\theta_x(x,\tau)$. Since

$$u_r(r,t) \leq 0 \ \text{ and } \ \theta_x(x,\tau) = y'(x) + \sum_{k=1}^{\infty} \tau^k y_k'(x),$$

as $\tau \to 0^+$ we have $y'(x) \leq 0$. At $r = 0$ we have $0 = u_r(0,t)$ which implies

$$0 = \theta_x(0,\tau) = y'(0) + \sum_{k=1}^{\infty} \tau^k y_k'(0).$$

As $\tau \to 0^+$ we have $y'(0) = 0$.

In addition, one can show that

$$\tau\theta_\tau = -[1 + \frac{xy'(x)}{2}] + \sum_{k=1}^{\infty} \tau^k [ky_k(x) - \frac{xy_k'(x)}{2}]$$

so that $-\tau\theta_\tau \to 1 + \frac{1}{2}xy'(x)$ as $\tau \to 0^+$. The formal (physical) argument is that the large temporal gradient in the singular region will be suppressed as the outer region is approached if

$$\frac{1}{2}xy'(x) + 1 \to 0 \ \text{ as } x \to \infty. \tag{3.33}$$

Thus, $y'(x) \to -\frac{2}{x}$ or $y(x) \sim -2\ln x + K$ as $x \to \infty$ and we expect the behavior

$$\theta(x,\tau) \sim -\ln\tau - 2\ln x + K$$

as $\tau \to 0^+$ so that for $r > 0$, the final time solution $U_F(r)$ behaves like $-2\ln r + K$ for some constant K.

In summary, the asymptotic behavior of $u(r,t)$ in a neighborhood of $(0,T)$ [or of $U_F(r)$ in a neighborhood of 0] is determined by the behavior of a solution $y(x)$ to

$$y'' + \left(\frac{n-1}{x} - \frac{x}{2}\right) y' + e^y - 1 = 0 \text{ and } y'(x) \le 0 \text{ for } x > 0$$

with boundary conditions

$$y(0) \ge 0, \quad y'(0) = 0, \quad \text{and} \quad \lim_{x\to\infty} [1 + \frac{1}{2} x y'(x)] = 0.$$

We will prove this boundary value problem has no solution for dimensions $n = 1$ or $n = 2$. This in turn implies that the formal final time analysis is incomplete.

In this section we will prove that the solution $u(r,t)$ of (3.23)-(3.24) satisfies the asymptotic condition

$$u(r,t) + \ln(T-t) \to 0 \text{ as } t \to T^-$$

uniformly on sets of the type $\{(r,t) : r \le C(T-t)^{1/2}\}$ for any constant $C \ge 0$. We also obtain a lower bound on $u(r,t)$ near $r = 0$ for t near the blow-up time T:

$$U_F(r) \ge \ln\left(\frac{2(n-2)}{r^2}\right)$$

for $r > 0$ sufficiently small.

We will need to use an estimate on $u_r(r,t)$ to determine certain bounds later on. For the radially symmetric case, these are easy to obtain; however, such bounds can be obtained even in the non-radially symmetric case by using the maximum principle. Since $u_t(r,t) \ge 0$ and $u_r(r,t) \le 0$, equation (3.23) gives us

$$0 \le u_t = u_{rr} + \frac{n-1}{r} u_r + f(u) \le u_{rr} + f(u).$$

Multiplying by u_r and integrating with respect to r (with lower limit of integration 0), we obtain

$$[u_r(r,t)]^2 \le [u_r(r,t)]^2 + 2I(u(r,t)) \le 2I(u(0,t)) \tag{3.34}$$

where $I(u) = \int_0^u f(s)\, ds$.

Although the problem considered by Kassoy, Poland, and Kapila involves the nonlinear term $f(u) = e^u$, the results about to be developed are adaptable to the nonlinearity $f(u) = u^p$ which we discussed in Section 3.3. Thus, we will give a parallel development using these two nonlinearities.

The Self-Similar Problem. To analyze the asymptotic behavior of $u(r,t)$, we make the following change of variables:

$$\sigma = \ln\left(\frac{T}{T-t}\right) \quad \text{and} \quad x = r(T-t)^{-1/2};$$

then the rectangle $(0,R)\times(0,T)$ transforms into the set

$$\Pi = \{(x,\sigma) : \sigma > 0,\ 0 < x < RT^{-1/2}e^{\sigma/2}\}.$$

If $f(u) = e^u$, then define

$$w(x,\sigma) = u(r,t) + \ln(T-t) \quad \text{and} \quad F(w) = e^w - 1;$$

if $f(u) = u^p$, then define

$$w(x,\sigma) = (T-t)^{\beta}u(r,t) \quad \text{and} \quad F(w) = w^p - \beta w$$

where $\beta = \frac{1}{p-1}$. The initial-boundary value problem (3.23)- (3.24) is transformed into

$$w_\sigma = w_{xx} + \left(\frac{n-1}{x} - \frac{x}{2}\right)w_x + F(w), \quad (x,\sigma)\in\Pi \tag{3.35}$$

where the initial-boundary conditions for the case $f(u) = e^u$ are

$$\begin{aligned} &w(x,0) = \phi(xT^{1/2}) + \ln T, \quad x \in (0, RT^{-1/2}) \\ &w_x(0,\sigma) = 0, \quad w(RT^{-1/2}e^{\sigma/2},\sigma) = -\sigma + \ln T, \quad \sigma\in(0,\infty) \end{aligned} \tag{3.36}$$

and the initial-boundary conditions for the case $f(u) = u^p$ are

$$\begin{aligned} &w(x,0) = T^{\beta}\phi(xT^{1/2}), \quad x\in(0,RT^{-1/2}) \\ &w_x(0,\sigma) = 0, \quad w(RT^{-1/2}e^{\sigma/2},\sigma) = 0, \quad \sigma\in(0,\infty). \end{aligned} \tag{3.37}$$

From Corollary 3.19 and Corollary 3.21, we have the following *a priori* bounds where $\xi > 0$ is sufficiently small:

$$0 \le w(0,\sigma) \le -\ln\xi, \quad \sigma \ge 0, \quad \text{for } f(u) = e^u \tag{3.38}$$

and

$$\beta^\beta \le w(0,\sigma) \le (\beta/\xi)^\beta, \quad \sigma\ge 0, \quad \text{for } f(u) = u^p. \tag{3.39}$$

Equation (3.34) implies the existence of a positive constant $\gamma = \gamma(\xi, f)$ such that

$$-\gamma \le w_x(x,\sigma) \le 0 \tag{3.40}$$

for all $(x,\sigma)\in\overline{\Pi}$. Combining this with (3.38) and (3.39) yields the estimate

$$-\gamma x \le w(x,\sigma) \le \mu \tag{3.41}$$

for some positive constant $\mu = \mu(\xi, f)$ and for all $(x, \sigma) \in \overline{\Pi}$. In fact, for $f(u) = u^p$, since $u \geq 0$, we have $w(x, \sigma) = (T-t)^\beta u(r, t) \geq 0$.

If there is a steady-state solution to (3.35)-(3.36) or (3.35)- (3.37), say $y(x) = \lim_{\sigma \to \infty} w(x, \sigma)$, then $y(x)$ must be a solution to equation (3.35), so

$$y'' + \left(\frac{n-1}{x} - \frac{x}{2}\right) y' + F(y) = 0, \quad x \in (0, \infty). \tag{3.42}$$

For the case $f(u) = e^u$, equations (3.36), (3.38), and (3.40) imply the conditions:

$$y(0) =: \alpha \geq 0, \quad y'(0) = 0, \quad \text{and} \quad -\infty < -\gamma \leq y'(x) \leq 0 \quad \text{for} \quad x \in (0, \infty). \tag{3.43}$$

For the case $f(u) = u^p$, equations (3.37), (3.39), (3.40), and $w \geq 0$ imply the conditions:

$$\begin{aligned} &y(0) =: \alpha \geq \beta^\beta, \quad y'(0) = 0, \quad \text{and} \\ &-\infty < -\gamma \leq y'(x) \leq 0 \quad \text{and} \quad y(x) \geq 0 \quad \text{for} \quad x \in (0, \infty). \end{aligned} \tag{3.44}$$

For $f(u) = e^u$ with $1 \leq n \leq 2$, the information above is enough to allow us to conclude that (3.42)-(3.43) has the unique solution $y(x) \equiv 0$. For $f(u) = u^p$ with $1 \leq n \leq 2$ or with $n > 2$ and $p \leq n/(n-2)$, the problem (3.42)-(3.44) has the unique solution $y(x) \equiv \beta^\beta$. This will be apparent in the proofs which are given later.

For $f(u) = e^u$ with $n > 2$, Eberly and Troy [EBE1] show that (3.42)-(3.43) has an infinite number of solutions (which in fact have the asymptotic property (3.33) which was conjectured by Kassoy, Poland, and Kapila). For $f(u) = u^p$ with $n > 2$ and $p > (n+2)/(n-2)$, Troy [TRO2] shows that (3.42)-(3.44) has an infinite number of solutions. By comparison of $w(x, \sigma)$ to a certain *singular solution* of equation (3.35), we will find that out of the infinite number of solutions to (3.42)-(3.43) or (3.42)-(3.44), the only steady-state solutions are $y(x) \equiv 0$ and $y(x) \equiv \beta^\beta$, respectively.

Behavior near Singular Solutions. For $f(u) = e^u$, the partial differential equation (3.35) has the singular solution

$$S_e(x) = \ln\left(\frac{2(n-2)}{x^2}\right)$$

for the cases $n > 2$. For $f(u) = u^p$, (3.35) has the singular solution

$$S_p(x) = \left(\frac{-4\beta\left(\beta + \frac{2-n}{2}\right)}{x^2}\right)^\beta$$

for the cases $n > 2$ and $p > \frac{n}{n-2}$. Note that

$$1 + \frac{1}{2} x S_e' = 0, \quad S_e'' + \frac{n-1}{x} S_e' + e^{S_e} = 0 \tag{3.45}$$

and

$$\beta S_p + \frac{1}{2} x S_p' = 0, \quad S_p'' + \frac{n-1}{x} S_p' + (S_p)^p = 0 \tag{3.46}$$

for $0 < x < \infty$.

We show that the singular solution must intersect the initial data $w(x,0)$. Consider first the singular solution $S_e(x)$; then

$$S_e(0^+) = \infty > w(0,0)$$

and

$$S_e(RT^{-1/2}) = \ln\left(\frac{2(n-2)T}{R^2}\right) < \ln T = w(RT^{-1/2}, 0)$$

since $2(n-2) < R^2$ for blowup in finite time (see Theorem 2.19). This proves that $w(x,0)$ intersects $S_e(x)$ at least once for $0 < x < RT^{-1/2}$.

For the singular solution $S_p(x)$, we can make the following observations:

$$S_p(0^+) = \infty > w(0,0) \text{ and } S_p(RT^{-1/2}) > 0 = w(RT^{-1/2}, 0).$$

If $w(x,0) \leq S_p(x)$ on the interval $[0, RT^{-1/2}]$, then by the maximum principle it must be that $w(x,\sigma) \leq S_p(x)$ on $\overline{\Pi}$. By the result of Troy [TRO2], any solution of (3.42)-(3.44) must intersect $S_p(x)$ transversally at least once. Thus, $y(x) = \lim_{\sigma\to\infty} w(x,\sigma) < S_p(x)$ for all $x > 0$. As we will see, the only solution of (3.42)-(3.44) which has this property is $y(x) \equiv 0$. Thus, $w(0,\sigma) \to 0$ as $\sigma \to \infty$, a contradiction to (3.39).

In either case, we conclude that there is a first $x_1 \in (0, RT^{-1/2})$ such that $w(x_1, 0) = S_*(x_1)$ and $w(x,0) < S_*(x)$ on $(0, x_1)$ for $* = e$ or p.

Lemma 3.22 *There is a continuously differentiable function $x_1(\sigma)$ with domain $[0,\infty)$ such that $x_1(0) = x_1$ and $w(x_1(\sigma),\sigma) = S_*(x_1(\sigma))$ for all $\sigma \geq 0$.*

Proof. Define $D(x,\sigma) = w(x,\sigma) - S_*(x)$. We claim that $\nabla D \neq 0$ whenever $D = 0$. By the maximum principle, $u_t(r,t) > 0$ on $(0,R) \times (0,T)$. Using the self-similar change of variables, we have

$$u_t = (T-t)^{-1}\left(w_\sigma + 1 + \tfrac{1}{2} x w_x\right) \text{ for } f(u) = e^u, \text{ and}$$
$$u_t = (T-t)^{-\beta-1}\left(w_\sigma + \beta w + \tfrac{1}{2} x w_x\right) \text{ for } f(u) = u^p.$$

If $\nabla D = 0$ at a point in Π where $D = 0$, then $D_\sigma = 0$ implies that $w_\sigma = 0$. The condition $D_x = 0$ implies

$$1 + \tfrac{1}{2} x w_x = 0 \text{ for } f(u) = e^u, \text{ and}$$
$$\beta w + \tfrac{1}{2} x w_x = 0 \text{ for } f(u) = u^p.$$

In either case, $u_t = 0$ is forced at some point in $(0,R) \times (0,T)$, a contradiction.

Secondly, we claim that $D_x \neq 0$ at any value $(\overline{x}, \overline{\sigma}) \in \Pi$ where $D(\overline{x}, \overline{\sigma}) = 0$ and $D(x, \overline{\sigma}) < 0$ in a left neighborhood of $\overline{x}$.

If $D(\overline{x}, \overline{\sigma}) = 0$ and $D_x(\overline{x}, \overline{\sigma}) = 0$, then equations (3.35), (3.45), and (3.46) imply that $D_{xx}(\overline{x}, \overline{\sigma}) = D_\sigma(\overline{x}, \overline{\sigma})$. In addition, since $u_t > 0$ we have $D_\sigma(\overline{x}, \overline{\sigma}) > 0$. Thus, $D_{xx}(\overline{x}, \overline{\sigma}) > 0$, which implies that $(\overline{x}, \overline{\sigma})$ is a local minimum point for D, a contradiction to $D < 0$ on a left neighborhood of $\overline{x}$. It must be that $D_x(\overline{x}, \overline{\sigma}) > 0$.

Recall that the initial data $u(r, 0) = \phi(r)$ satisfies the inequality $\Delta\phi + f(\phi) \geq 0$. This implies that

$$D_{xx}(x, 0) + \frac{n-1}{x} D_x(x, 0) + F(w(x, 0)) - F(S_*(x)) \geq 0.$$

On a left neighborhood of x_1, this in turn yields

$$\left(x^{n-1} D_x(x, 0)\right)_x \geq 0.$$

An integration yields $D_x(x_1, 0) > 0$. By the implicit function theorem, there is a continuously differentiable function $x_1(\sigma)$ such that $x_1(0) = x_1$ and $D_x(x_1(\sigma), \sigma) = 0$ for some maximal interval $[0, \sigma_0)$. If $\sigma_0 < \infty$, then by continuity, $D(x_1(\sigma_0), \sigma_0) = 0$. But $D_x(x_1(\sigma_0), \sigma_0) > 0$, so the implicit function theorem allows an extension of the domain past σ_0, a contradiction to the maximality of $[0, \sigma_0)$. Thus, $\sigma_0 = \infty$. □

For $f(u) = u^p$, since $w(0, 0) < S_p(0^+)$, $w(RT^{-1/2}, 0) < S_p(RT^{-1/2})$, and $w(x_1, 0) = S_p(x_1)$ transversally, there must be a last point of intersection between $w(x, 0)$ and $S_p(x)$, say $x_L \in (x_1, RT^{-1/2})$. A construction similar to Lemma 3.22 leads to the existence of a continuously differentiable function $x_L(\sigma)$ with domain $[0, \infty)$ such that $x_L(0) = x_L$ and $w(x_L(\sigma), \sigma) = S_p(x_L(\sigma))$ for $\sigma \geq 0$.

Let $\Pi_1 = \{(x, \sigma) : \sigma > 0,\ 0 < x < x_1(\sigma)\}$. We can prove the following comparison result on this set.

Lemma 3.23 *The function $D(x, \sigma) = w(x, \sigma) - S_*(x)$ satisfies $D(x, \sigma) < 0$ for $(x, \sigma) \in \Pi_1$.*

Proof. By Lemma 3.22, we have shown that $D \leq 0$ on the parabolic boundary of Π_1. Since $F(w)$ is a locally one-sided Lipschitz continuous function, we can apply the maximum principle to obtain $D \leq 0$ on $\overline{\Pi}_1$.

If $D(x_0, \sigma_0) = 0$ for some $(x_0, \sigma_0) \in \Pi_1$, then $\nabla D(x_0, \sigma_0) = 0$ since $D \leq 0$ on Π_1. But we had shown in Lemma 3.22 that $\nabla D \neq 0$ whenever $D = 0$. Thus, it must be that $D(x, \sigma) < 0$ for $(x, \sigma) \in \Pi_1$. □

Define the value

$$x_2 = \sup\{x \in (x_1, RT^{-1/2}] : D(s, 0) \geq 0 \text{ for } s \in [x_1, x]\}.$$

Since $D(x_1, 0) = 0$ and $D_x(x_1, 0) > 0$, the supremum exists. For $f(u) = e^u$ we have $x_2 \leq RT^{-1/2}$ and for $f(u) = u^p$ we have $x_2 \leq x_L < RT^{-1/2}$. Define $x_2(\sigma) = x_2 e^{\sigma/2}$ and define the set

$$\Pi_2 = \{(x, \sigma) : \sigma > 0,\ x_1(\sigma) < x < x_2(\sigma)\}.$$

Lemma 3.24 *The function $D(x, \sigma) = w(x, \sigma) - S_*(x)$ satisfies $D(x, \sigma) > 0$ for $(x, \sigma) \in \Pi_2$.*

Proof. Let $E(\sigma) = D(x_2(\sigma), \sigma)$. By definition of x_2, $E(0) = D(x_2, 0) \geq 0$. Also,

$$E'(\sigma) = D_\sigma(x_2(\sigma), \sigma) + \frac{1}{2} x_2(\sigma) D_x(x_2(\sigma), \sigma).$$

As in Lemma 3.22, using $u_t(r, t) \geq 0$, we have

$$E'(\sigma) \geq 0 \text{ for } f(u) = e^u, \text{ and}$$
$$e^{-\beta\sigma} \tfrac{d}{d\sigma} \left[e^{\beta\sigma} E(\sigma)\right] = E'(\sigma) + \beta E(\sigma) \geq 0 \text{ for } f(u) = u^p.$$

In either case, an integration yields $E(\sigma) \geq 0$ for $\sigma \geq 0$.

On the parabolic boundary of Π_2 we now have $D \geq 0$. By the maximum principle, $D \geq 0$ on $\overline{\Pi}_2$. An argument similar to the one used in Lemma 3.23 shows that $D(x, \sigma) > 0$ for $(x, \sigma) \in \Pi_2$. □

Corollary 3.25 *For each $N > 0$ there is a $\sigma_N > 0$ such that for each $\sigma > \sigma_N$, $w(x, \sigma)$ intersects $S_*(x)$ at most once for $x \in [0, N]$.*

Proof. For each $N > 0$ let σ_N be the solution to $N = x_2 \exp(\frac{1}{2}\sigma_N)$. Lemma 3.23 guarantees that $D(x, \sigma) < 0$ for $x \in [0, x_1(\sigma))$ and Lemma 3.24 guarantees that $D(x, \sigma) > 0$ for $x \in (x_1(\sigma), x_2(\sigma)]$. For $\sigma > \sigma_N$ we have $[0, N] \subseteq [0, x_2(\sigma)]$ by definition of σ_N, so $D(\cdot, \sigma) = 0$ at most once on this interval. □

In a later subsection on the convergence results, we will see that $x_1(\sigma) \to L$ as $\sigma \to \infty$ where $S_e(L) = 0$ or $S_p(L) = \beta^\beta$.

Analysis of the Steady-State Problem. In this subsection we will analyze the behavior of the boundary value problems (3.42)-(3.43) and (3.42)-(3.44) which we restate here:

$$y'' + \left(\frac{n-1}{x} - \frac{x}{2}\right) y' + F(y) = 0, \quad x \in (0, \infty).$$

The boundary conditions for $f(u) = e^u$ are

$$y(0) = \alpha \geq 0, \quad y'(0) = 0, \quad \text{and} \quad -\gamma \leq y'(x) \leq 0 \text{ for } x \in (0, \infty),$$

and the boundary conditions for $f(u) = u^p$ are

$$y(0) = \alpha \geq \beta^\beta, \quad y'(0) = 0, \quad \text{and} \quad -\gamma \leq y'(x) \leq 0, y(x) \geq 0 \text{ for } x \in (0, \infty).$$

Lemma 3.26 *Consider the initial-value problem associated with (3.42).*

1. *Any solution to (3.42)-(3.43) must satisfy* $y(\sqrt{2n}\,) \leq 0$.
2. *Any solution to (3.42)-(3.44) must satisfy* $y(\sqrt{2n}\,) \leq \beta^\beta$.

Proof. To prove part (1), $F(y) = e^y - 1 \geq y$, so equation (3.42) implies that

$$y'' + \left(\frac{n-1}{x} - \frac{x}{2}\right) y' + y \leq 0$$

for any solution $y(x)$ of (3.42). Let $u(x) = \alpha(1 - \frac{x^2}{2n})$; then

$$u'' + \left(\frac{n-1}{x} - \frac{x}{2}\right) u' + u = 0, \quad u(0) = y(0), \quad \text{and} \quad u'(0) = y'(0).$$

Define $W(x) = u(x)y'(x) - u'(x)y(x)$. While $u(x) > 0$,

$$W' + \left(\frac{n-1}{x} - \frac{x}{2}\right) W \leq 0 \text{ and } W(0) = 0,$$

so an integration yields $W(x) \leq 0$. But

$$\left(\frac{y}{u}\right)'(x) = \frac{W(x)}{[u(x)]^2} \leq 0,$$

so integrating from 0 to $\sqrt{2n}$ yields $y(\sqrt{2n}\,) \leq u(\sqrt{2n}\,) = 0$.

Note that for $\alpha > 0$, if $y(z) = 0$, then $y'(z) < 0$ by uniqueness to initial value problems, so $y(x) < 0$ for $x > z$.

To prove part (2), $F(y) = y^p - \beta y$ is convex, so $F(y) \geq y - \beta^\beta$ and equation (3.42) implies that

$$v'' + \left(\frac{n-1}{x} - \frac{x}{2}\right) v' + v \leq 0$$

where $v(x) = y(x) - \beta^\beta$ and $y(x)$ is a solution of (3.42). A similar argument as for part (1) shows that $v(\sqrt{2n}\,) \leq 0$ and so $y(\sqrt{2n}\,) \leq \beta^\beta$.

Note that for $\alpha > \beta^\beta$, if $y(z) = \beta^\beta$, then $y'(z) < 0$ by uniqueness to initial value problems, so $y(x) < \beta^\beta$ for $x > z$. □

Define the function $h(x)$ by

$$h(x) = y''(x) + \frac{n-1}{x} y'(x).$$

Define the function $g(x)$ by

$$g(x) = 1 + \tfrac{1}{2} x y'(x) \text{ for } F(y) = e^y - 1, \text{ and}$$
$$g(x) = \beta y(x) + \tfrac{1}{2} x y'(x) \text{ for } F(y) = y^p - \beta y.$$

It can be shown that h and g satisfy the following equations:

$$\begin{aligned} &g'' + \left(\tfrac{n-1}{x} - \tfrac{x}{2}\right) g' + [F'(y) - 1]g = 0, \\ &g(0) > 0, \quad g'(0) = 0, \end{aligned} \tag{3.47}$$

$$\begin{aligned} &h'' + \left(\tfrac{n-1}{x} - \tfrac{x}{2}\right) h' + [F'(y) - 1]h = -F''(y)[y']^2, \\ &h(0) \leq 0, \quad h'(0) = 0, \end{aligned} \tag{3.48}$$

$$g' - \frac{1}{2}xg = -\frac{1}{2}xe^y + \frac{2-n}{2}y' \text{ for } F(y) = e^y - 1, \text{ and} \tag{3.49}$$

$$g' - \frac{1}{2}xg = -\frac{1}{2}xy^p + \left[\beta + \frac{2-n}{2}\right] y' \text{ for } F(y) = y^p - \beta y. \tag{3.50}$$

In addition, define $W(x) = g(x)h'(x) - g'(x)h(x)$; then

$$W' + \left(\frac{n-1}{x} - \frac{x}{2}\right) W = -F''(y)[y']^2 g, \text{ and } W(0) = 0.$$

An integration yields

$$\begin{aligned} W(x) &= -x^{1-n}e^{x^2/4} \int_0^x s^{n-1}e^{-s^2/4}F''\left(y(x)\right)\left[y'(x)\right]^2 g(x)\, ds \\ &=: -x^{1-n}e^{x^2/4} I(x) \end{aligned}$$

where $I(x) \geq 0$ while $g(x) > 0$. Note that $(h/g)'(x) = W(x)/[g(x)]^2$, so while $g(x) > 0$ we have

$$h(x) = \frac{h(0)}{g(0)}g(x) - g(x)\int_0^x t^{1-n}e^{t^2/4}\frac{I(t)}{[g(t)]^2}\, dt. \tag{3.51}$$

Lemma 3.27 *Consider the initial value problem for (3.42).*

1. *If $y(x)$ is a solution to (3.42)-(3.43) with $\alpha > 0$, then $g(x)$ must have a zero.*

2. *If $y(x)$ is a solution to (3.42)-(3.44) with $\alpha > \beta^\beta$, then $g(x)$ must have a zero.*

Proof. Suppose that $g(x) \geq \varepsilon > 0$ for all $x \geq 0$. Equation (3.51) implies that

$$h(x) \leq \frac{h(0)}{g(0)}g(x) \leq -\delta < 0$$

since $h(0)/g(0) < 0$ and since $I(x) \geq 0$. Multiplying by x^{n-1} and integrating yields

$$y'(x) \leq -\frac{\delta}{n}x \text{ for } x \geq 0$$

which contradicts the boundedness of $y'(x)$ assumed in (3.43) and (3.44). Thus, $g(x)$ cannot be bounded away from zero.

Suppose that $g(x) > 0$ for $x \geq 0$ and that g is not bounded away from zero. Suppose there is an increasing unbounded sequence $\{x_k\}_{k=1}^{\infty}$ such that $g'(x_k) = 0$. Equation (3.47) implies that

$$g''(x_k) = [1 - F'(y(x_k))]\, g(x_k).$$

But Lemma 3.26 implies that

$$1 - F'(y(x_k)) > 0$$

for k sufficiently large, which forces $g''(x_k) > 0$ for k large. This is a contradiction since g would have two local minimums without a local maximum between. It must be that $g'(x) < 0$ for x sufficiently large and $g(x) \to 0$ as $x \to \infty$.

Suppose there is an increasing unbounded sequence $\{x_k\}_{k=1}^{\infty}$ such that $g''(x_k) = 0$ and $g'(x_k) \leq -L < 0$. Equation (3.47) implies that

$$\left(\frac{n-1}{x_k} - \frac{x_k}{2}\right) g'(x_k) + [F'(y(x_k)) - 1] g(x_k) = 0$$

where $g'(x_k) \leq -L$, $F'(y(x_k)) - 1$ is bounded, and $g(x_k) \to 0$. The left-hand side of the last equality must become infinite, a contradiction. Thus, $g'(x) < 0$ for x large and $g'(x) \to 0$.

In (3.48), take the limit as $x \to \infty$ to obtain

$$\begin{aligned}
\lim_{x \to \infty} h(x) &= -\lim_{x \to \infty} g(x) \textstyle\int_0^x t^{1-n} e^{t^2/4} \frac{I(t)}{[g(t)]^2}\, dt \\
&= \lim_{x \to \infty} x^{1-n} e^{x^2/4} \frac{I(x)}{g'(x)} \\
&= -\infty
\end{aligned}$$

where we have used L'Hôspital's rule. This implies that $h(x) \leq -\delta < 0$ for x sufficiently large. Multiplying by x^{n-1} and integrating yields

$$y'(x) \leq K - \frac{\delta}{n} x$$

for some constant K and for x large. As before, this contradicts the boundedness of $y'(x)$ assumed in (3.43) and (3.44).

In all of the above cases we arrived at contradictions, so there must be a value x_0 such that $g(x_0) = 0$ and $g(x) > 0$ for $x \in [0, x_0)$. □

Lemma 3.28 *Consider the problem (3.42)-(3.43).*

1. *If $1 \leq n \leq 2$, then the only solution is $y(x) \equiv 0$.*
2. *If $n > 2$, then the only solution which intersects $S_e(x)$ exactly once is $y(x) \equiv 0$.*

Proof. Let $1 \leq n \leq 2$; then $\frac{2-n}{2} \geq 0$. Suppose there is an $\alpha > 0$ for which (3.42)-(3.43) has a solution. Let x_0 be the first zero for $g(x)$. Suppose there is an $x_1 > x_0$ such that $g'(x_1) = 0$ and $g(x) < 0$ on $(x_0, x_1]$. Equation (3.49) implies that

$$\begin{aligned} 0 &< -\tfrac{1}{2}x_1 g(x_1) = g'(x_1) - \tfrac{1}{2}x_1 g(x_1) \\ &= -\tfrac{1}{2}x_1 e^{y(x_1)} + \tfrac{2-n}{2} y'(x_1) \\ &< 0 \end{aligned}$$

which is a contradiction. Thus, $g'(x) < 0$ for $x \geq x_0$ and so $g(x) \leq -\varepsilon < 0$ for $x \geq \overline{x} > x_0$. But

$$h(x) = g(x) - e^{y(x)} \leq g(x) \leq -\varepsilon.$$

Multiplying by x^{n-1} and integrating yields

$$y'(x) \leq K - \frac{\varepsilon}{n} x,$$

contradicting the boundedness of $y'(x)$. As a result, the only solution of (3.42)-(3.43) for $1 \leq n \leq 2$ is $y(x) \equiv 0$.

Let $n > 2$. Define $D(x) = y(x) - S_e(x)$; then

$$\begin{aligned} &D'' + \left(\tfrac{n-1}{x} - \tfrac{x}{2}\right) D' + \tfrac{2(n-2)}{x^2}(e^D - 1) = 0, \quad x \in (0, \infty), \\ &D(0^+) = -\infty, \quad D'(0^+) = \infty. \end{aligned} \tag{3.52}$$

Note that $D'(x) > 0$ while $D(x) < 0$ for x in a right neighborhood of 0. Suppose that $D(x) < 0$ for all $x \geq 0$; then

$$e^D - 1 < 0 \text{ and } D'' + \left(\frac{n-1}{x} - \frac{x}{2}\right) D' \geq 0.$$

Integrating yields

$$x^{n-1} e^{-x^2/4} D'(x) \geq \overline{x}^{n-1} e^{-\overline{x}^2/4} D'(\overline{x}) =: p > 0, \quad \text{for } x \geq \overline{x}.$$

Consequently,

$$D(x) \geq D(\overline{x}) + p \int_{\overline{x}}^{x} t^{1-n} e^{t^2/4} dt.$$

But the right-hand side of this inequality must be positive for x large, contradicting our assumption that $D < 0$. Thus, $D(x)$ must have a first zero x_1 and $D'(x) > 0$ on $(0, x_1]$.

By Lemma 3.27, $g(x)$ must have a first zero x_0. But then

$$D'(x_0) = \frac{2}{x_0} g(x_0) = 0 \text{ and } x_0 > x_1.$$

If $D(x_0) < 0$, then there must have been a second zero x_2 for D. Otherwise, $D(x) > 0$ on $(x_1, x_0]$. Suppose that $D > 0$ for all $x \geq x_0$; then there is an $\overline{x}$ sufficiently large such that

$$D(\overline{x}) > 0, \quad D'(\overline{x}) < 0, \quad D''(\overline{x}) > 0, \quad \text{and} \quad \frac{n-1}{\overline{x}} - \frac{\overline{x}}{2} < 0.$$

Evaluating equation (3.52) at $\overline{x}$ yields

$$0 < D''(\overline{x}) + \left(\frac{n-1}{\overline{x}} - \frac{\overline{x}}{2}\right) D'(\overline{x}) + \frac{2(n-2)}{\overline{x}^2}\left(e^{D(\overline{x})} - 1\right) = 0,$$

a contradiction. Thus, D must have a second zero x_2.

We have shown that there are at least two points of intersection between the graphs of $y(x)$ and $S_e(x)$ for $\alpha > 0$. Thus, the only solution to (3.42)-(3.43) which intersects $S_e(x)$ exactly once is $y(x) \equiv 0$. □

Lemma 3.29 *Consider the problem (3.42)-(3.44).*

1. *If $1 \leq n \leq 2$, or, if $n > 2$ and $\beta + \frac{2-n}{2} \geq 0$, then the only solution is $y(x) \equiv \beta^\beta$.*
2. *If $n > 2$ and $\beta + \frac{2-n}{2} < 0$, then the only solution which intersects $S_p(x)$ exactly once is $y(x) \equiv \beta^\beta$.*

Proof. To prove part (1), suppose there is an $\alpha > \beta^\beta$ such that (3.42)-(3.44) has a solution. Let x_0 be the first zero for $g(x)$. Suppose there is an $x_1 > x_0$ such that $g'(x_1) = 0$ and $g(x) < 0$ on $(x_0, x_1]$. Equation (3.49) implies that

$$\begin{aligned} 0 &< -\tfrac{1}{2}x_1 g(x_1) \\ &= g'(x_1) - \tfrac{1}{2}x_1 g(x_1) \\ &= -\tfrac{1}{2}x_1[y(x_1)]^p + \left[\beta + \tfrac{2-n}{2}\right] y'(x_1) \\ &\leq 0 \end{aligned}$$

which is a contradiction. Thus, $g'(x_0) < 0$ for $x \geq x_0$ and so by an argument similar to that in Lemma 3.28, we obtain $y'(x) \leq K - \frac{\varepsilon}{n}x$, a contradiction to the boundedness of y'. That is, the only solution to (3.42)-(3.44) for part (1) is $y(x) \equiv 0$.

To prove part (2), assume $\alpha > \beta^\beta$ and let $y(x)$ be the solution to (3.42)-(3.44). Define $W(x) = y(x)S_p'(x) - y'(x)S_p(x)$ and $Q(u) = F(u)/u$; then

$$W' + \left(\frac{n-1}{x} - \frac{x}{2}\right) W = yS_p[Q(y) - Q(S_p)]. \tag{3.53}$$

Note that $Q(u)$ is an increasing function. Also note that

$$W(x) = -2Kx^{-2\beta-1}g(x) \tag{3.54}$$

where $S_p(x) = Kx^{-2\beta}$. Since $n - 2 - 2\beta > 0$, we have $x^{n-1}W(x) \to 0$ as $x \to 0^+$. Integrating (3.53) we obtain

$$x^{n-1}e^{-x^2/4}W(x) = \int_0^x t^{n-1}e^{-t^2/4}y(t)S_p(t)\left[Q\left(y(t)\right) - Q\left(S_p(t)\right)\right] dt.$$

If $0 < y < S_p$ for all $x \geq 0$, then since $Q(u)$ is increasing, $W(x) < 0$ for all x. But then (3.54) implies $g(x) > 0$ for all x, a contradiction to Lemma 3.27. Consequently, there must be a value z such that $y(z) = S_p(z)$.

Also, $W(x) < 0$ for $x \in [0, x_0)$ where x_0 is the first zero of $g(x)$. At x_0, $0 < W'(x_0)$ which implies that $y(x_0) > S_p(x_0)$, where we have used (3.53) and the fact that Q is increasing. Note that $W'(z) \neq 0$ since $y(x)$ and $S_p(x)$ are two linearly independent solutions to the same differential equation. Thus, $z < x_0$ is necessary.

Let $x_1 > x_0$ be small enough so that $W(x_1) > 0$. Suppose that $y > S_p$ for all $x > z$, then $Q(y) > Q(S_p)$ and an integration of (3.53) yields

$$x^{n-1}e^{-x^2/4}W(x) \geq x_1^{n-1}e^{-x_1^2/4}W(x_1) =: p > 0.$$

But $(S_p/y)'(x) = W(x)/[y(x)]^2$, so

$$\frac{S_p}{y}(x) \geq \frac{S_p}{y}(x_1) + p\int_{x_1}^x t^{1-n}e^{t^2/4}[y(t)]^{-2}\, dt.$$

For x sufficiently large, the right-hand side must become larger than 1, in which case $(S_p/y)(x) \geq 1$, a contradiction to our assumption that $y > S_p$ for $x > z$. Therefore, there is another value q where $y(q) = S_p(q)$.

We have shown that there are at least two points of intersection between the graphs of $y(x)$ and $S_p(x)$ for $\alpha > \beta^\beta$. Thus, the only solution of (3.42)-(3.44) which intersects $S_p(x)$ exactly once is $y(x) \equiv \beta^\beta$. □

The Convergence Results. We are now able to precisely describe how the blowup asymptotically evolves for (3.23)-(3.24) by looking at the self-similar problems (3.35)-(3.36) or (3.35)-(3.37).

Theorem 3.30 *Consider the partial differential equation (3.35).*

1. *The solution $w(x, \sigma)$ to (3.35)-(3.36) converges to 0 as $\sigma \to \infty$ uniformly in x on compact subsets of $[0, \infty)$.*

2. *The solution $w(x, \sigma)$ of (3.35)-(3.37) converges to β^β as $\sigma \to \infty$ uniformly in x on compact subsets of $[0, \infty)$.*

Proof. Define $w^m(x, \sigma) := w(x, \sigma + m)$ for $m \geq 0$. We will show that as $m \to \infty$, $w^m(x, \sigma)$ converges to the solution $y(x)$ of (3.42)-(3.43) or (3.42)-(3.44) uniformly on compact subsets of $\mathbb{R}^+ \times \mathbb{R}$. As long as the limiting function is unique, it is equivalent to prove that given any unbounded increasing sequence $\{n_j\}$, there exists a subsequence [renamed] $\{n_j\}$ such that w^{n_j} converges to $y(x)$ uniformly on compact subsets of $\mathbb{R}^+ \times \mathbb{R}$.

Let $N \in \mathbb{N}$. For i sufficiently large, the rectangle given by

$$Q_{2N} = \{(x, \sigma) : 0 \leq x \leq 2N, \ |\sigma| \leq 2N\}$$

lies in the domain of w^{n_i}. The radially symmetric function $\tilde{w}(\varsigma, \sigma) = w^{n_i}(|\varsigma|, \sigma)$ solves the parabolic equation

$$\tilde{w}_\sigma = \Delta \tilde{w} - \frac{1}{2}\langle \varsigma, \nabla \tilde{w} \rangle + F(\tilde{w})$$

on the cylinder given by

$$\Gamma_{2N} = \{(\varsigma, \sigma) : \mathbb{R}^n \times \mathbb{R} : |\varsigma| \leq 2N, \ |\sigma| \leq 2N\}$$

with $-2N\gamma \leq \tilde{w}(\varsigma, \sigma) \leq \mu$ using (3.41).

By Schauder's interior estimates, all partial derivatives of $\tilde{w}$ can be uniformly bounded on the subcylinder $\Gamma_N \subset \Gamma_{2N}$. Consequently, w^{n_i}, $w^{n_i}_\sigma$, and $w^{n_i}_{xx}$ are uniformly Lipschitz continuous on $Q_N \subset Q_{2N}$ and their Lipschitz constants depend on N but not on i. By the Arzela-Ascoli Theorem, there is a subsequence $\{n_j\}$ and a function $\overline{w}$ such that w^{n_j}, $w^{n_j}_\sigma$, and $w^{n_j}_{xx}$ converge to $\overline{w}$, $\overline{w}_\sigma$, and $\overline{w}_{xx}$, respectively, uniformly on Q_N.

Repeating the construction for all N and taking a diagonal subsequence, we can conclude that

$$w^{n_j} \to \overline{w}, \quad w^{n_j}_\sigma \to \overline{w}_\sigma, \quad \text{and} \quad w^{n_j}_{xx} \to \overline{w}_{xx}$$

uniformly on every compact subset in $\mathbb{R}^+ \times \mathbb{R}$. Clearly $\overline{w}$ satisfies (3.35)-(3.36)-(3.38)-(3.40) or (3.35)- (3.37)-(3.39)-(3.40). For $n > 2$ and $F(w) = e^w - 1$, or, for $n > 2$ with $\beta + (2-n)/2 < 0$ and $F(w) = w^p - \beta w$, the function $\overline{w}$ intersects $S_e(x)$ at most once since, by Corollary 3.25, $w^{n_j}(x, \sigma)$ intersects $S_e(x)$ at most once on $[0, N]$ for each $\sigma > \sigma_N$.

We now prove that $\overline{w}$ is independent of σ. For the solution $w(x, \sigma)$ of (3.35)-(3.36) or (3.35)-(3.37), define the energy functional

$$E(\sigma) = \int_0^v \rho(x) \left[\frac{1}{2}w_x^2 - G(w)\right] dx \tag{3.55}$$

where $v = RT^{-1/2}e^{\sigma/2}$, $\rho(x) = x^{n-1}e^{-x^2/4}$, and where

$$G(w) = e^w - w \ \text{ for } \ F(w) = e^w - 1, \ \text{ and}$$
$$G(w) = \tfrac{1}{p+1}w^{p+1} - \tfrac{1}{2}\beta w^2 \ \text{ for } \ F(w) = w^p - \beta w.$$

Multiplying equation (3.35) by ρw_σ and integrating from 0 to v yields the equation

$$\begin{aligned} \textstyle\int_0^v \rho w_\sigma^2 \, dx &= \textstyle\int_0^v w_\sigma (\rho w_x)_x \, dx + \int_0^v [\rho G(w)]_\sigma \, dx \\ &= \textstyle\int_0^v \left[\rho G(w) - \frac{1}{2}\rho w_x^2\right]_\sigma dx + \rho w_\sigma w_x|_{x=0}^{x=v}. \end{aligned} \tag{3.56}$$

Moreover,

$$E'(\sigma) = \int_0^v \frac{\partial}{\partial\sigma}\left[\tfrac{1}{2}\rho w_x^2 - \rho G(w)\right] dx + \tfrac{1}{2} v\rho(v)\left[\tfrac{1}{2} w_x^2(v,\sigma) - G(w(v,\sigma))\right]. \tag{3.57}$$

Let a and b be chosen so that $0 \le a < b$. Integrate equation (3.56) with respect to σ from a to b and use (3.57) to obtain

$$\begin{aligned}\int_a^b \int_0^v \rho w_\sigma^2 \, d\sigma &= -\int_a^b E'(\sigma)\, d\sigma + \int_a^b \rho(v) w_\sigma(v,\sigma) w_x(v,\sigma)\, d\sigma \\ &\quad + \tfrac{1}{2}\int_a^b v\rho(v)\left[\tfrac{1}{2} w_x^2(v,\sigma) - G(w(v,\sigma))\right] d\sigma \\ &=: E(a) - E(b) + \psi(a,b).\end{aligned} \tag{3.58}$$

Recall that $|w_x| \le \gamma$ and observe that

$$w_\sigma(v,\sigma) = -1 - Ru_r(R, T(1-e^{-\sigma})) \ \text{ for } \ f(u) = e^u, \ \text{ and,}$$
$$w_\sigma(v,\sigma) = -Ru_r(R, T(1-e^{-\sigma})) \ \text{ for } \ f(u) = u^p.$$

We see that in either case w_σ is uniformly bounded as $\sigma \to \infty$. We conclude that

$$\lim_{a\to\infty}\left(\sup_{b>a}\psi(a,b)\right) = 0. \tag{3.59}$$

For any fixed N we will prove that

$$\iint_{Q_N} \rho \overline{w}_\sigma^2 \, dx\, d\sigma = \lim_{n_j\to\infty} \iint_{Q_N} \rho\left(w_\sigma^{n_j}\right)^2 dx\, d\sigma = 0.$$

It is not a restriction to assume that $\lim_{j\to\infty}(n_{j+1} - n_j) = \infty$. For all j sufficiently large,

$$N \le RT^{-1/2}\exp\left(\frac{1}{2}(n_j - N)\right) \ \text{ and } \ n_{j+1} - n_j \ge 2N.$$

Consequently,

$$\begin{aligned}&\int_{-N}^N \int_0^N \rho\left(w_\sigma^{n_j}\right)^2 dx\, d\sigma \\ &\le \int_{-N}^{-N+n_{j+1}-n_j} \int_0^{RT^{-1/2}\exp(\frac{\sigma+n_j}{2})} \rho\left(w_\sigma^{n_j}\right)^2 dx\, d\sigma \\ &= E(n_j - N) - E(n_{j+1} - N) + \psi(n_j - N, n_{j+1} - N)\end{aligned}$$

where we have used (3.58). Applying equation (3.59) gives us

$$\iint_{Q_N} \rho\overline{w}_\sigma^2 \, dx\, d\sigma \le \limsup_{j\to\infty}[E(n_j - N) - E(n_{j+1} - N)]. \tag{3.60}$$

Choose any K arbitrarily large. For j sufficiently large we have

$$
\begin{aligned}
&E(n_j - N) - E(n_{j+1} - N) = \\
&\textstyle\int_0^K \frac{1}{2}\rho \left[(w_x^{n_j}(x,-N))^2 - (w_x^{n_{j+1}}(x,-N))^2 \right] dx \\
&- \textstyle\int_0^K \rho \left[G(w^{n_j}(x,-N)) - G(w^{n_{j+1}}(x,-N)) \right] dx \\
&+ \textstyle\int_K^{\frac{R}{\sqrt{T}} e^{\frac{n_j - N}{2}}} \rho \left[\frac{1}{2} (w_x^{n_j}(x,-N))^2 - G(w^{n_j}(x,-N)) \right] dx \\
&- \textstyle\int_K^{\frac{R}{\sqrt{T}} e^{\frac{n_{j+1} - N}{2}}} \rho \left[\frac{1}{2} (w_x^{n_{j+1}}(x,-N))^2 - G(w^{n_{j+1}}(x,-N)) \right] dx.
\end{aligned}
\tag{3.61}
$$

The first two integrals in (3.61) converge to zero as $j \to \infty$. The boundedness of $w_x^{n_j}$ and w^{n_j} imply that the absolute value of the last two integrals in (3.61) is bounded by

$$M \int_K^\infty x^{n-1} \exp\left(-\frac{1}{4}x^2\right) dx$$

where M is a positive constant. This integral can be made arbitrarily small by choosing K sufficiently large.

Thus, we have proved that $\int_{-N}^{N} \rho \overline{w}_\sigma^2 \, dx \, d\sigma = 0$ for all N, which in turn implies $\overline{w}_\sigma = 0$. We have

$$\overline{w}(x,\sigma) = \overline{w}(x,0) = y(x)$$

where $y(x)$ is a nonincreasing globally Lipschitz continuous solution of (3.42)-[(3.43) or (3.44)]. For the cases where there is a singular solution $S_*(x)$, the function $y(x)$ intersects $S_*(x)$ exactly once on $[0,\infty)$ since $\overline{w}$ does. By Lemma 3.28, the only possibility for $y(x)$ is $y(x) \equiv 0$ in the case $f(u) = e^u$. By Lemma 3.29, the only possibility for $y(x)$ is $y(x) \equiv \beta^\beta$ in the case $f(u) = u^p$. □

Corollary 3.31 *Let $u(r,t)$ be the solution to (3.23)-(3.24). In the case $f(u) = e^u$,*

$$u(r,t) + \ln(T-t) \to 0 \quad \textit{as} \quad t \to T^-$$

uniformly for $r \le C(T-t)^{1/2}$ for arbitrary $C \ge 0$. In particular,

$$\lim_{t \to T^-} [u(0,t) + \ln(T-t)] = 0.$$

In the case $f(u) = u^p$,

$$(T-t)^\beta u(r,t) \to \beta^\beta \quad \textit{as} \quad t \to T^-$$

uniformly for $r \le C(T-t)^{-1/2}$ for arbitrary $C \ge 0$. In particular,

$$\lim_{t \to T^-} (T-t)^\beta u(0,t) = \beta^\beta.$$

Corollary 3.32 *Consider IBVP (3.23)-(3.24) for the cases where the singular solution $S_*(r)$ exists. There is a value $r_1 \in (0, R)$ such that the following properties are valid:*

1. $u(r_1, 0) = S_*(r_1)$,
2. $u(r, 0) < S_*(r)$ *for* $0 < r < r_1$, *and*
3. *for each* $r \in (0, r_1)$ *there is a* $\bar{t} = \bar{t}(r) \in (0, T)$ *such that* $u(r, t) > S_*(r)$ *for* $t \in (\bar{t}, T)$.

Proof. Theorem 3.30 guarantees that the first branch of zeros $x_1(\sigma)$ of $D(x, \sigma) = w(x, \sigma) - S_*(x)$ is bounded and converges to the number ℓ where $S_e(\ell) = 0$ or $S_p(\ell) = \beta^\beta$.

Define $r_1 = x_1 T^{1/2}$; then $D(x_1, 0) = 0$ implies that $u(r_1, 0) = S_*(r_1)$. In addition, $u(r, 0) < S_*(r)$ for $r \in (0, r_1)$.

Since $x_1(\sigma)$ is bounded and since $\frac{d}{d\sigma} D(rT^{-1/2} e^{\sigma/2}, \sigma) \geq 0$ for each $r \in (0, r_1)$, there is a value $\overline{\sigma} > 0$ such that

$$rT^{-1/2} e^{\overline{\sigma}/2} = x_1(\overline{\sigma}), \quad D(x_1(\overline{\sigma}), \overline{\sigma}) = 0, \quad \text{and,} \quad D(rT^{-1/2} e^{\sigma/2}, \sigma) > 0$$

for all $\sigma > \overline{\sigma}$. Changing back to the variables (r, t) with $\overline{\sigma} = \ln[T/(T - \bar{t})]$ gives us $u(r, t) > S_*(r)$ for $t \in (\bar{t}, T)$. □

3.5 Comments

The basic fundamental theory for parabolic problems can be found in [FRI1], [LAD], and [HEN]. The existence results for parabolic initial boundary value problems assuming the existence of upper and lower solutions are an outgrowth of similar results for elliptic problems. These in turn can be traced back to the Perron method for solving the Dirichlet problem (see [GIL]). Sattinger [SAT1] used a monotone iteration scheme to prove Theorem 3.1 assuming that $f(x, t, u)$ is C^1 with respect to u. A "Perron-type" proof of Theorem 3.1 is given in [BEB2] assuming only Hölder continuity, but the existence should be viewed as a consequence of invariance as discussed in Chapter 4.

The problem of nonexistence for parabolic problems waas driven by the need to understand supercritical thermal events. The first systematic development of the supercritical case for the ignition problem is due to Kassoy, Liñan, and co-workers [KAS1],[KAS2],[KAS3],[KAS4] but either for extremely simplified models or the results were numerical in nature.

An important paper by Ball [BAL] discussed the ideas of nonexistence and blowup. The earliest blowup results are due to Kaplan [KPL] and Fujita [FUJ1], but the first to study blowup for the ignition model is the paper by Bebernes and Kassoy [BEB4]. In that paper, upper and lower bounds for the blowup time are rigorously proved. The best results on when blowup occurs are due to Lacey [LAC1] and Bellout [BEL].

In Section 3.3, the results presented are essentially due to Friedman and McLeod [FRI2]. Numerical experimentation [KAS3],[KAP1] clearly predicted that, for radially symmetric domains, as the blowup time was approached a single hot spot developed. For $f(u) = u^p$, p sufficiently large, and for sufficiently large initial data, Weissler [WES3] was the first to prove single point blowup. Friedman and McLeod [FRI2] extended these results significantly using a clever maximum principle argument.

The question of where blowup occurs can be answered by giving a characterization of the asymptotic behavior of the solution near blowup. Kassoy and Poland [KAS3] and Kapila [KAP1] independently had predicted that the final time solution profile would be of the form $-2\ln r$ at the blowup point. Dold [DOL] challenged their prediction and suggested a different variable grouping to describe the thermal runaway process. The Kassoy-Kapila prediction in part based on numerics has since been proved incorrect [BEB9], [EBE1],[EBE2],[FRI4],[TRO2]. One can describe the asymptotics of the solution in a backward space-time parabola. This was first done by Giga and Kohn [GIG5] for $f(u) = u^p$ with $p < 1 + 2/n$ and has been extended to more general nonlinearities in [GIG6],[BEB10], [BEB11]. The ideas of using backward similarity variables can be traced back to Leray [LER] in 1934.

Determining the final time solution profile at time $t = T$ remains a difficult and unsolved problem. Using the Dold [DOL] similarity grouping, several numerical studies have been conducted [BER],[GAL1] which support the conjectures of Dold and of Galaktionov and Posashkov [GAL2].

4

The Complete Model for Solid Fuel

In this chapter we discuss comparison techniques, invariant sets, and existence results related to invariance. Our main application is the complete solid fuel model (1.24)-(1.25):

$$\left.\begin{aligned} T_t - \Delta T &= \varepsilon\delta y^m \exp\left(\tfrac{T-1}{\varepsilon T}\right) \\ y_t - \beta\Delta y &= -\varepsilon\delta\Gamma y^m \exp\left(\tfrac{T-1}{\varepsilon T}\right) \end{aligned}\right. , \quad (x,t) \in \Omega \times (0,\infty)$$

with initial-boundary conditions

$$\begin{aligned} &T(x,0) = 1, \; y(x,0) = 1, \quad x \in \Omega \\ &T(x,t) = 1, \; \tfrac{\partial y(x,t)}{\partial \eta(x)} = 0, \quad (x,t) \in \partial\Omega \times (0,\infty). \end{aligned}$$

where $\beta \geq 0$, $\Gamma > 0$, and $\delta > 0$. We prove there is a solution (T,y) for all $(x,t) \in \Omega \times (0,\infty)$ such that $y(x,t) \to 0$ as $t \to \infty$.

Section 4.1 covers comparison techniques. These methods are generalizations of maximum principles. The comparisons for systems of equations require the concept of a quasimonotone function.

In Section 4.2 we discuss invariance results. The main idea is that of an invariant set Σ which contains the range of solutions to a given initial-boundary value problem. The results rely on a geometric concept of outer normals to $\partial\Sigma$ which in some sense prevent solutions from exiting Σ.

The existence results of Section 4.3 are closely related to the invariance results. The main tool used in proving existence of solutions is the Leray-Schauder degree theory.

4.1 Comparison Techniques

We begin with the scalar inequalities which we used in Chapter 3. The following result is a generalization of the maximum principle.

Theorem 4.1 *Let* $\Pi_T = \Omega \times (0,T)$ *and* $\Gamma_T = (\Omega \times \{0\}) \cup (\partial\Omega \times (0,T])$. *Suppose* $u, v \in C(\overline{\Pi}_t, \mathbb{R}) \cap C^{2,1}(\Pi_T, \mathbb{R})$ *are two functions such that*

$$\begin{aligned} &u_t - \Delta u - f(x,t,u) < v_t - \Delta v - f(x,t,v), \quad (x,t) \in \Pi_T \\ &u(x,t) < v(x,t), \quad (x,t) \in \Gamma_T; \end{aligned}$$

then $u(x,t) < v(x,t)$ *for all* $(x,t) \in \overline{\Pi}_T$.

Proof. Let $d = v - u$. Assume that the conclusion is false; then there is a first $\bar{t} > 0$ such that $d(\bar{x}, \bar{t}) = 0$ for some $\bar{x} \in \Omega$, $d(x,t) > 0$ for $(x,t) \in \overline{\Omega} \times [0, \bar{t})$, and $d_t(\bar{x}, \bar{t}) \leq 0$. Moreover, $d(x, \bar{t})$ attains its minimum at $x = \bar{x}$, so $\nabla d(\bar{x}, \bar{t}) = 0$ and $\Delta d(\bar{x}, \bar{t}) \geq 0$. However,

$$d_t(\bar{x}, \bar{t}) = v_t - u_t > \Delta v - \Delta u - f(\bar{x}, \bar{t}, v) + f(\bar{x}, \bar{t}, u) \geq 0,$$

a contradiction. Thus, $u(x,t) < v(x,t)$ for all $(x,t) \in \overline{\Pi}_T$. □

Corollary 4.2 *If* $f(x,t,u+z) - f(x,t,u) \leq Lz$ *for* $0 < z < \delta$ *and if* $u, v \in C(\overline{\Pi}_T, \mathbb{R}) \cap C^{2,1}(\Pi_T, \mathbb{R})$ *have the properties*

$$\begin{aligned} & u_t - \Delta u - f(x,t,u) \leq v_t - \Delta v - f(x,t,v), \quad (x,t) \in \Pi_T \\ & u(x,t) \leq v(x,t), \quad (x,t) \in \Gamma_T, \end{aligned}$$

then $u(x,t) \leq v(x,t)$ *for all* $(x,t) \in \overline{\Pi}_T$.

Proof. Set $v^\varepsilon = v + \varepsilon e^{2Lt}$ for $\varepsilon > 0$; then

$$\begin{aligned} v_t^\varepsilon - \Delta v^\varepsilon - f(x,t,v^\varepsilon) &= v_t + 2L\varepsilon e^{2Lt} - \Delta v - f(x,t,v+\varepsilon e^{2Lt}) \\ &\geq v_t + 2L\varepsilon e^{2Lt} - \Delta v - f(x,t,v) - L\varepsilon e^{2Lt} \\ &= v_t - \Delta v - f(x,t,v) + L\varepsilon e^{2Lt} \\ &> v_t - \Delta v - f(x,t,v) \\ &\geq u_t - \Delta u - f(x,t,u) \end{aligned}$$

for $(x,t) \in \Pi_T$, and where we have used the one-sided Lipschitz condition for $f(x,t,u)$. Moreover,

$$v^\varepsilon(x,t) = v(x,t) + \varepsilon e^{2Lt} > v(x,t) \geq u(x,t)$$

for $(x,t) \in \Gamma_T$. The hypotheses of Theorem 4.1 are satisfied for the functions v^ε and u. Consequently, $v^\varepsilon(x,t) > u(x,t)$ for all $(x,t) \in \overline{\Pi}_T$. This inequality and $v(x,t) = \lim_{\varepsilon \to 0^+} v^\varepsilon(x,t)$ imply $v(x,t) \geq u(x,t)$ for all $(x,t) \in \overline{\Pi}_T$. □

Theorem 3.2 is a restatement of Theorem 4.1 and Corollary 4.2. These results do not extend to systems unless we impose an extra condition. For $a, b \in \mathbb{R}^n$, define the orderings "$<$" and "$\leq$" by

$$a < b \;\Leftrightarrow\; a_i < b_i \;\forall i, \;\text{ and, }\; a \leq b \;\Leftrightarrow\; a_i \leq b_i \;\forall i.$$

Definition 4.1 *A function $f : D \subset \mathbb{R}^n \to \mathbb{R}^n$ is* quasimonotone nondecreasing *if f_i is nondecreasing in x_j for $j \neq i$. That is, if $x \leq y$ and $x_i = y_i$ for $x, y \in D$, then $f_i(x) \leq f_i(y)$ for all $i = 1, \ldots, n$. Similar definitions can be given for quasimonotone nonincreasing, quasimonotone decreasing, and quasimonotone increasing.*

Consider the parabolic system

$$\frac{\partial u_k}{\partial t} - \alpha_k \Delta u_k = f_k(x, t, u), \quad k = 1, \ldots, n,$$

where $u = (u_1, \ldots, u_n)$. For $\alpha = (\alpha_1, \ldots, \alpha_n)$ and $f = (f_1, \ldots, f_n)$, the system of equations may be written in the form

$$u_t - \alpha \bullet \Delta u = f(x, t, u, \nabla u).$$

Theorem 4.3 *Let $u, v \in C(\overline{\Pi}_T, \mathbb{R}^n) \cap C^{2,1}(\Pi_T, \mathbb{R}^n)$ be functions which satisfy the conditions*

$$u_t - \alpha \bullet \Delta u - f(x, t, u) < v_t - \alpha \bullet \Delta v - f(x, t, v), \quad (x, t) \in \Pi_T$$
$$u(x, t) < v(x, t), \quad (x, t) \in \Gamma_T.$$

If f is quasimonotone nondecreasing, then $u(x, t) < v(x, t)$ for all $(x, t) \in \overline{\Pi}_T$.

The proof of Theorem 4.3 goes essentially as in the 1-dimensional case.

Initial-boundary value problem (1.24)-(1.25) is not quasimonotone, so Theorem 4.3 is not applicable. There is a very useful comparison theorem, however.

For $z : \Pi_T \to \mathbb{R}^m$, define $L_i z_i = \frac{\partial z_i}{\partial t} - E_i z_i$ where E_i is a uniformly elliptic operator with uniformly bounded coefficients on $\overline{\Pi}_T$ and is of the form

$$E_i = \sum_{j,k} a_{ijk} \frac{\partial^2}{\partial x_j \partial x_k} + \sum_j bij \frac{\partial}{\partial x_j}.$$

In the following development, we consider functions $F, \underline{F} : \overline{\Pi}_T \times \mathbb{R}^n \to \mathbb{R}^n$ such that $F(x, t, u) \geq \underline{F}(x, t, u)$, $\underline{F}$ is Lipschitz continuous in u uniformly in (x, t), and $\underline{F}$ is quasimonotone increasing in u.

For $u, v \in \mathbb{R}^n$, define the *replacement vector* $\hat{u}^j(v)$ by

$$\hat{u}^j(v) = (u_1, \ldots, u_{j-1}, v_j, u_{j+1}, \ldots, u_n).$$

The next result is due to Fife [FIF].

Theorem 4.4 *Let $u, \underline{u} \in C(\overline{\Pi}_T, \mathbb{R}^n) \cap C^{2,1}(\Pi_T, \mathbb{R}^n)$ satisfy*

$$Lu - F(x, t, u) \geq 0 \geq L\underline{u} - \underline{F}(x, t, \underline{u}), \quad (x, t) \in \Pi_T$$
$$u(x, 0) \geq \underline{u}(x, 0), \quad x \in \overline{\Omega},$$

and either

$$u(x,t) \geq \underline{u}(x,t) \quad or \quad \frac{\partial u}{\partial \eta}(x,t) \geq \frac{\partial \underline{u}}{\partial \eta}(x,t), \quad (x,t) \in \partial\Omega \times (0,T);$$

then $u(x,t) \geq \underline{u}(x,t)$ *for all* $(x,t) \in \overline{\Pi}_T$.

Proof. For $\varepsilon > 0$, define $v = \underline{u} - \frac{\varepsilon}{\sqrt{n}}(1+2\ell t)(1,\ldots,1)$ where ℓ is the Lipschitz constant for $\underline{F}$; then

$$\begin{aligned} L_i v_i &= L_i \underline{u}_i - 2\varepsilon\ell \\ &\leq \underline{F}_i(x,t,\underline{u}) - 2\varepsilon\ell \\ &\leq \underline{F}_i(x,t,v) + \varepsilon\ell(1+2\ell t) - 2\varepsilon\ell \\ &= \underline{F}_i(x,t,v) + \varepsilon\ell(2\ell t - 1) \end{aligned}$$

where the Lipschitz continuity of $\underline{F}$ was used. Thus, for $t \in [0, \frac{\ell}{2}]$ we have $L_i v_i \leq \underline{F}_i(x,t,v)$.

Set $w = u - v$. For each j we have

$$\begin{aligned} L_j w_j &= L_j u_j - L_j v_j \\ &\geq F_j(x,t,u) - \underline{F}_j(x,t,v) \\ &\geq \underline{F}_j(x,t,u) - \underline{F}_j(x,t,v) \end{aligned} \tag{4.1}$$

for $(x,t) \in \Omega \times (0, \frac{\ell}{2}]$. Since $w(x,0) \geq \frac{\varepsilon}{\sqrt{n}}(1,\ldots,1)$, there is a $\tau \in (0, \frac{\ell}{2}]$ such that $w \geq 0$ on $[0,\tau)$.

Since $\underline{F}_j$ is nondecreasing in u_i for $i \neq j$, we have

$$\begin{aligned} \underline{F}_j(x,t,u) &= \underline{F}_j(x,t,\hat{u}^j(u)) \\ &\geq \underline{F}_j(x,t,\hat{v}^j(u)) \\ &\geq \underline{F}_j(x,t,v) - \ell(u_j - v_j) \end{aligned} \tag{4.2}$$

for $t \in [0,\tau]$.

Combining the inequalities (4.1) and (4.2) yields $L_j w_j + \ell w_j \geq 0$ for $(x,t) \in \Omega \times (0,\tau)$ with $w_j(x,0) > 0$ for $x \in \Omega$ and either $w_j(x,t) \geq 0$ or $\frac{\partial w_j(x,t)}{\partial \eta} \geq 0$ for $(x,t) \in \partial\Omega \times (0,\tau)$. By the maximum principle, $w_j(x,t) > 0$ on $\Omega \times [0,\tau]$ for each $j = 1,\ldots,n$. Thus, $w(x,t) > 0$ on $\Omega \times [0, \frac{\ell}{2}]$.

Let $\varepsilon \to 0^+$; then $u(x,t) \geq \underline{u}(x,t)$ on $\overline{\Omega} \times [0, \frac{\ell}{2}]$. The arguments above may be repeated on intervals of the type $[\frac{k\ell}{2}, \frac{(k+1)\ell}{2}]$ for $k \geq 1$ to eventually obtain the result on $\overline{\Pi}_T$.. □

Similar results hold for functions F and $\overline{F}$ where $\overline{F}$ is Lipschitz continuous in u uniformly in (x,t) and where $\overline{F}$ is quasimonotone nonincreasing.

We need only require that for $u, \overline{u} \in C(\overline{\Pi}_T, \mathbb{R}^n) \cap C^{2,1}(\Pi_T, \mathbb{R}^n)$:

$$\begin{aligned} &L\overline{u} - \overline{F}(x,t,\overline{u}) \geq 0 \geq Lu - F(x,t,u), \quad (x,t) \in \Pi_T \\ &\overline{u}(x,0) \geq u(x,0), \quad x \in \overline{\Omega} \end{aligned}$$

and either

$$\overline{u}(x,t) \geq u(x,t) \text{ or } \frac{\partial \overline{u}}{\partial \eta}(x,t) \geq \frac{\partial u}{\partial \eta}(x,t), \quad (x,t) \in \partial\Omega \times (0,T);$$

to guarantee that $\overline{u}(x,t) \geq u(x,t)$ for all $(x,t) \in \overline{\Pi}_T$.

The application of these comparison results to the solid fuel model is as follows. Let $T = 1 + \varepsilon\theta$ in equations (1.24)-(1.25). The system may be written as

$$\begin{aligned} &\left.\begin{aligned} &\theta_t - \Delta\theta = \delta y^m \exp\left(\tfrac{\theta}{1+\varepsilon\theta}\right) \\ &y_t - \beta\Delta y = -\gamma y^m \exp\left(\tfrac{\theta}{1+\varepsilon\theta}\right) \end{aligned}\right\}, \quad (x,t) \in \Pi_T \\ &\theta(x,0) = 0, \ y(x,0) = 1, \quad x \in \overline{\Omega} \\ &\theta(x,t) = 0, \ \tfrac{\partial y(x,t)}{\partial \eta} = 0, \quad (x,t) \in \partial\Omega \times (0,T) \end{aligned}$$

where $\gamma = \varepsilon\delta\Gamma$. Since $\theta \geq 0$, $1 \leq \exp(\frac{\theta}{1+\varepsilon\theta}) \leq \exp\left(\frac{1}{\varepsilon}\right)$ for $\varepsilon > 0$. Moreover, since $0 \leq y \leq 1$, we have

$$\delta y^m \leq \delta y^m \exp\left(\frac{\theta}{1+\varepsilon\theta}\right) \leq \delta y^m \exp(\frac{1}{\varepsilon})$$

and

$$-\gamma y^m \exp\left(\frac{1}{\varepsilon}\right) \leq -\gamma y^m \exp\left(\frac{\theta}{1+\varepsilon\theta}\right) \leq -\gamma y^m.$$

Consider the comparison systems

$$\begin{aligned} &\left.\begin{aligned} &\overline{\theta}_t - \Delta\overline{\theta} = \delta \overline{y}^m \exp\left(\tfrac{1}{\varepsilon}\right) \\ &\overline{y}_t - \beta\Delta\overline{y} = -\gamma\overline{y}^m \end{aligned}\right\}, \quad (x,t) \in \Pi_T \\ &\overline{\theta}(x,0) = 0, \ \overline{y}(x,0) = 1, \quad x \in \overline{\Omega} \\ &\overline{\theta}(x,t) = 0, \ \tfrac{\partial \overline{y}(x,t)}{\partial \eta} = 0, \quad (x,t) \in \partial\Omega \times (0,T) \end{aligned} \tag{4.3}$$

and

$$\begin{aligned} &\left.\begin{aligned} &\underline{\theta}_t - \Delta\underline{\theta} = \delta \underline{y}^m \\ &\underline{y}_t - \beta\Delta\underline{y} = -\gamma\underline{y}^m \exp\left(\tfrac{1}{\varepsilon}\right) \end{aligned}\right\}, \quad (x,t) \in \Pi_T \\ &\underline{\theta}(x,0) = 0, \ \underline{y}(x,0) = 1, \quad x \in \overline{\Omega} \\ &\underline{\theta}(x,t) = 0, \ \tfrac{\partial \underline{y}(x,t)}{\partial \eta} = 0, \quad (x,t) \in \partial\Omega \times (0,T). \end{aligned} \tag{4.4}$$

Both system (4.3) and system (4.4) are quasimonotone nondecreasing. By Theorem 4.4 we have the comparisons

$$(\underline{\theta}, 0) \leq (\underline{\theta}, \underline{Y}) \leq (\theta, Y) \leq (\overline{\theta}, \overline{Y}) \leq (\overline{\theta}, 1).$$

where $(\overline{\theta}, \overline{y})$ is a solution to (4.3) and $(\underline{\theta}, \underline{y})$ is a solution to (4.4). Note that

$$\overline{y} = \overline{y}(t) = \left\{ \begin{array}{ll} e^{-\gamma t}, & m = 1 \\ \left(\frac{1}{1+\gamma(m-1)t}\right)^{\frac{1}{m-1}}, & m > 1 \end{array} \right\}$$

and

$$\underline{y} = \underline{y}(t) = \left\{ \begin{array}{ll} e^{-\gamma \exp(1/\varepsilon)t}, & m = 1 \\ \left(\frac{1}{1+\gamma(m-1)\exp(1/\varepsilon)t}\right)^{\frac{1}{m-1}}, & m > 1 \end{array} \right\},$$

so $\underline{y} \to 0$ and $\overline{y} \to 0$ as $t \to \infty$. Since $\underline{y} \leq y \leq \overline{y}$, we have $y(x,t) \to 0$ as $t \to \infty$.

To construct useful majorant and minorant reaction functions $\overline{F}$ and $\underline{F}$, define

$$\underline{F}_j(x,t,u) = \inf\{F_j(x,t,\tilde{v}^j(u)) : v \geq u\}$$

and

$$\overline{F}_j(x,t,u) = \sup\{F_j(x,t,\tilde{v}^j(u)) : v \leq u\}.$$

When u is increased, the infimum is taken over a smaller set and the supremum is taken over a larger set. Consequently, $\underline{F}_j$ and $\overline{F}_j$ are nondecreasing. If F is Lipschitz continuous, then so are $\underline{F}$ and $\overline{F}$.

Corollary 4.5 *Let $\underline{u}, \overline{u} \in \mathbb{R}^n$ where $\underline{u} < \overline{u}$. Suppose that F satisfies*

$$F_j(x,t,\tilde{u}^j(\underline{u})) \geq 0 \geq F_j(x,t,\tilde{u}^j(\overline{u}))$$

for $\underline{u}_k \leq u_k \leq \overline{u}_k$ $(k \neq j)$ and for $(x,t) \in \Pi_T$.

Let u be a solution of

$$\begin{aligned} &u_t - Eu = F(x,t,u), \quad (x,t) \in \Omega \times (0,T) \\ &\underline{u} \leq u(x,0) \leq \overline{u}, \quad x \in \Omega \end{aligned} \tag{4.5}$$

where E is a uniformly elliptic operator, and with either

$$\underline{u} \leq u(x,t) \leq \overline{u}, \quad (x,t) \in \partial\Omega \times (0,T) \tag{4.6}$$

or

$$\frac{\partial u}{\partial \eta}(x,t) = 0, \quad (x,t) \in \partial\Omega \times (0,T); \tag{4.7}$$

then $\underline{u} \leq u(x,t) \leq \overline{u}$ for $(x,t) \in \overline{\Pi}_T$.

Proof. Construct $\underline{F}$ and $\overline{F}$ by defining

$$\underline{F}_j = \inf\{F_j(x,t,\tilde{v}^j(u)) : u \le v \le \overline{u}\}$$

and

$$\overline{F}_j = \sup\{F_j(x,t,\tilde{v}^j(u)) : \underline{u} \le v \le u\}.$$

Theorem 4.4 can be applied using these values for $\underline{F}$ and $\overline{F}$. □

Corollary 4.6 *Let $u(x,t)$ be a solution to (4.5)-(4.6) or -(4.7). Suppose that $\underline{F}$ and $\overline{F}$ are independent of x. Let $\alpha(t)$ be a solution to*

$$\alpha' = \underline{F}(t,\alpha), \quad \alpha(0) = \underline{u}$$

and let β be a solution to

$$\beta' = \overline{F}(t,\beta), \quad \beta(0) = \overline{u}$$

with $a \le \underline{u} < \overline{u} \le b$ for some $a, b \in \mathbb{R}^n$; then $\alpha(t) \le u(x,t) \le \beta(t)$ for all $(x,t) \in \overline{\Pi}_T$.

4.2 Invariance

Let $\Omega \subset \mathbb{R}^n$ be a bounded domain with boundary $\partial\Omega$ of class $C^{2+\alpha}$ for some $\alpha \in (0,1)$. For $T > 0$, define the cylinder $\Pi_T = \Omega \times (0,T)$ with lateral boundary $S_T = \partial\Omega \times (0,T)$. For $u \in C^{2,1}(\Pi_T, \mathbb{R}^m)$, define $Eu = (Eu_1, \ldots, Eu_m)$ by

$$E(x,t)u_k(x,t) = \sum_{i,j} a_{ij}(x,t) \frac{\partial^2 u_k(x,t)}{\partial x_i \partial x_j}$$

where $a_{ij} \in C^{\alpha,\alpha/2}(\overline{\Pi}_T, \mathbb{R})$. We assume that the matrix $A = [a_{ij}]$ is symmetric and positive definite with uniformly bounded coefficients, so E is a uniformly elliptic operator.

Let $M(x,t,u) = [M_{ij}(x,t,u)]$ be an $m \times m$ matrix where

$$M_{ij} \in C^{\alpha,\alpha/2,\alpha}(\overline{\Pi}_T \times \mathbb{R}^m, \mathbb{R}^m).$$

We can choose M so that the operators

$$\frac{\partial}{\partial t}(\cdot)_\ell - \sum_{i,j,k} M_{\ell k} a_{ij} \frac{\partial^2 (\cdot)_k}{\partial x_i \partial x_j}, \quad \ell = 1, \ldots, m,$$

are uniformly parabolic.

Let $f(x,t,u,p) \in C^{\alpha,\alpha/2,\alpha,\alpha}(\overline{\Pi}_T \times \mathbb{R}^m \times \mathbb{R}^{nm}, \mathbb{R}^m)$ where $p = [p_{ij}]$ is an $n \times m$ matrix.

In this section we will consider the quasilinear parabolic initial value problem

$$u_t = M(x,t,u)E(x,t)u + f(x,t,u,Du), \quad (x,t) \in \Pi_T \tag{4.8}$$
$$u(x,0) = \phi(x), \quad x \in \overline{\Omega} \tag{4.9}$$

where $\phi \in C^{2+\alpha}(\overline{\Omega}, \mathbb{R}^m)$ and $Du = [\partial u_j/\partial x_i]_{n\times m}$, with either of the boundary conditions

$$u(x,t) = \psi(x,t) \in C^{2+\alpha}(\overline{S}_T, \mathbb{R}^m) \tag{4.10}$$

or

$$\frac{\partial u(x,t)}{\partial \eta(x)} = h(x,t,u(x,t)) \in C^{1+\alpha,1+\alpha/2,1+\alpha}(\overline{S}_T \times \mathbb{R}^m, \mathbb{R}^m) \tag{4.11}$$

where $\eta(x)$ is an outward normal vector to $\partial\Omega$ at x. Assume that for the Dirichlet boundary condition, ϕ and ψ satisfy the compatability conditions

$$\psi(x,0) = \phi(x)$$

and

$$\psi_t(x,0) = M(x,0,\psi(x))E(x,0)\phi(x) + f(x,0,\phi(x),D\phi(x))$$

for $x \in \partial\Omega$. Assume that for the Neumann-type boundary conditions, ϕ and h satisfy the compatability condition

$$\frac{\partial \phi(x)}{\partial \eta(x)} = h(x,0,\phi(x))$$

for $x \in \partial\Omega$.

Definition 4.2 *The set $\Sigma \subset \mathbb{R}^m$ is* positively invariant *relative to*

1. *IBVP (4.8)-(4.9)-(4.10) if any solution u satisfies $u(\overline{\Pi}_T) \subset \Sigma$ provided $\phi \in C^{2+\alpha}(\overline{\Omega}, \Sigma)$ and $\psi \in C^{2+\alpha}(\overline{S}_T, \Sigma)$, or*
2. *IBVP (4.8)-(4.9)-(4.11) if any solution u satisfies $u(\overline{\Pi}_T) \subset \Sigma$ provided $\phi \in C^{2+\alpha}(\overline{\Omega}, \Sigma)$.*

Definition 4.3 *The matrix $M(x,t,u)$ satisfies the* eigenvalue condition *on Σ if for each $(x,t) \in \Pi_T$ and $u \in \partial\Sigma$ there exists an outer normal vector $N(x,t,u)$ to $\partial\Sigma$ at u, and a $\lambda(x,t,u) > 0$, such that*

$$N(x,t,u)^T[M(x,t,u) - \lambda(x,t,u)I] = 0. \tag{4.12}$$

This conditions says that N is a left eigenvector of M corresponding to the eigenvalue λ.

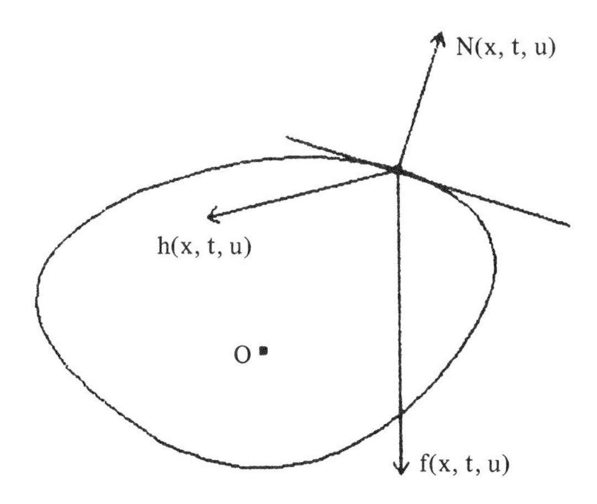

Figure 4.1.

Definition 4.4 *Consider IBVP (4.8)-(4.9)-(4.10). The function f satisfies the* strong flux condition *on Σ if for each $u \in \partial\Sigma$ there is an outer normal vector $N(x,t,u)$ to $\partial\Sigma$ at u such that*

$$N(x,t,u) \bullet f(x,t,u,P) < 0 \tag{4.13}$$

for those matrices P such that $PN(x,t,u) = 0$ for $(x,t) \in \overline{\Pi}_T$.

Definition 4.5 *Consider IBVP (4.8)-(4.9)-(4.11). The function f satisfies the* strong flux condition *on Σ if for each $u \in \partial\Sigma$ there is an outer normal vector $N(x,t,u)$ to $\partial\Sigma$ at u such that*

$$\begin{aligned} &N(x,t,u) \bullet f(x,t,u,P) < 0 \ \textit{for those matrices } P \textit{ such that} \\ &PN(x,t,u) = 0 \ \textit{for} \ (x,t) \in \overline{\Pi}_T, \ \textit{and} \\ &P^T \eta(x) = h(x,t,u) \ \textit{for} \ (x,t) \in \overline{S}_T, \end{aligned} \tag{4.14}$$

and

$$N(x,t,u) \bullet h(x,t,u) \leq 0 \ \textit{for} \ (x,t) \in \overline{S}_T. \tag{4.15}$$

In either definition, if instead $N \bullet f \leq 0$, then f is said to satisfy the *weak flux condition.*

These last definitions are a geometric concept which is illustrated in Figure 4.1.

We will show in the next two theorems that the strong flux condition will force solutions u to (4.8) to remain in Σ whenever the initial and boundary data are in Σ.

Theorem 4.7 *Let $\Sigma \subset \mathbb{R}^m$ be an open bounded convex set which contains 0. Let the matrix M satisfy the eigenvalue condition (4.12), and let f satisfy the strong flux condition (4.13), with the same outer normal vector N. If $\phi \in C^{2+\alpha}(\overline{\Omega}, \Sigma)$ and $\psi \in C^{2+\alpha}(\overline{S}_T, \Sigma)$, then Σ is positively invariant relative to IBVP (4.8)-(4.9)-(4.10).*

Proof. We prove the theorem by contradiction. Suppose that $u \in C(\overline{\Pi}_T) \cap C^{2,1}(\Pi_T)$ is a solution to (4.8)-(4.9)-(4.10) whose values do not remain in Σ. There must be a first time $t_0 \in (0,T]$ and some $x_0 \in \Omega$ such that $u_0 = u(x_0, t_0) \in \partial\Sigma$ and $u(x,t) \in \Sigma$ for all $(x,t) \in \overline{\Omega} \times [0, t_0)$.

Since f satisfies the strong flux condition, there is a normal vector $N_0 = N(x_0, t_0, u_0)$ such that $N_0 \bullet f(x_0, t_0, u_0, P) < 0$ for matrices P such that $PN_0 = 0$. Define $w(x,t) = [u(x,t) - u_0] \bullet N_0$. By the construction of t_0, $w(x_0, t_0) = 0$, $w(x,t) \leq 0$ for $(x,t) \in \overline{\Omega} \times [0, t_0]$, and $w_t(x_0, t_0) \geq 0$. Therefore, $w(\cdot, t_0)$ attains its maximum at x_0, so $\nabla w(x_0, t_0) = 0$ and the matrix $[w_{x_i x_j}]$ is negative semidefinite. Consequently, $E(x_0, t_0)w(x_0, t_0) = \sum_{i,j} a_{ij}(x_0, t_0) w_{x_i x_j}(x_0, t_0) \leq 0$.

The condition $\nabla w(x_0, t_0) = 0$ implies $Du(x_0, t_0)N_0 = 0$, so $Du_0 := Du(x_0, t_0)$ is a matrix such that $N_0 \bullet f(x_0, t_0, u_0, Du_0) < 0$. Also,

$$N_0^T M(x_0, t_0, u_0) = \lambda(x_0, t_0, u_0) N_0^T$$

by the eigenvalue condition. Thus,

$$\begin{aligned} w_t(x_0, t_0) &= u_t(x_0, t_0) \bullet N_0 \\ &= N_0^T [M(x_0, t_0, u_0) E(x_0, t_0) u(x_0, t_0) + f(x_0, t_0, u_0, Du_0)] \\ &= \lambda(x_0, t_0, u_0) N_0^T E(x_0, t_0) u(x_0, t_0) + N_0 \bullet f(x_0, t_0, u_0, Du_0) \\ &= \lambda(x_0, t_0, u_0) E(x_0, t_0) w(x_0, t_0) + N_0 \bullet f(x_0, t_0, u_0, Du_0) \\ &< 0 \end{aligned}$$

since $E(x_0, t_0)w(x_0, t_0) \leq 0$ and $N_0 \bullet f(x_0, t_0, u_0, Du_0) < 0$. This is a contradiction to $w_t(x_0, t_0) \geq 0$. Thus, Σ is positively invariant. □

Theorem 4.8 *Let $\Sigma \subset \mathbb{R}^m$ be an open bounded convex set which contains 0. Let the matrix M satisfy the eigenvalue condition (4.12), and let f satisfy the strong flux condition (4.14)-(4.15), with the same outer normal vector N. If $\phi \in C^{2+\alpha}(\overline{\Omega}, \Sigma)$, then Σ is positively invariant relative to IBVP (4.8)-(4.9)- (4.11).*

Proof. This theorem is also proved by contradiction. Let $u \in C(\overline{\Pi}_T) \cap C^{2,1}(\Pi_T)$ be a solution to (4.8)-(4.9)-(4.11) whose values do not remain in Σ. There must be a first time $t_0 \in (0,T]$ and some $x_0 \in \overline{\Omega}$ such that $u_0 = u(x_0, t_0) \in \partial\Sigma$ and $u(x,t) \in \Sigma$ for all $(x,t) \in \overline{\Omega} \times [0, t_0)$.

Since f satisfies the strong flux condition, there is a normal vector $N_0 = N(x_0, t_0, u_0)$ such that $N_0 \bullet f(x_0, t_0, u_0, P) < 0$ for matrices P such that $PN_0 = 0$. Define $w(x,t) = [u(x,t) - u_0] \bullet N_0$.

If $x_0 \in \Omega$, then as in Theorem 4.7, $w(\cdot, t_0)$ has a maximum of 0 at x_0, $\nabla w(x_0, t_0) = 0$, $w_t(x_0, t_0) \geq 0$, and $E(x_0, t_0)w(x_0, t_0) \leq 0$. Since $Du_0 N_0 = \nabla w(x_0, t_0) = 0$, where $Du_0 = Du(x_0, t_0)$, we have $N_0 \bullet f(x_0, t_0, u_0, Du_0) < 0$ by (4.14). Moreover, $w_t(x_0, t_0) < 0$ follows in a similar fashion as in Theorem 4.7, a contradiction to $w_t(x_0, t_0) \geq 0$.

Suppose $x_0 \in \partial\Omega$. The matrix $P = Du_0$ satisfies the conditions $PN_0 = 0$ and $P^T \eta(x_0) = h(x_0, t_0, u_0)$ since

$$h(x_0, t_0, u_0) = \frac{\partial u(x_0, t_0)}{\partial \eta(x_0)} = (Du_0)^T \eta(x_0)$$

by definition of the normal derivative. By (4.14), $N_0 \bullet f(x_0, t_0, u_0, Du_0) < 0$. By continuity of f and w_t, and by the piecewise smoothness of $\partial\Sigma$, there is an open ball B with center (x_0, t_0) such that

$$E(x, t_0)w(x, t_0) = \frac{w_t(x,t_0) - N(x,t_0,u(x,t_0)) \bullet f(x,t_0,u(x,t_0),Du(x,t_0))}{\lambda(x,t_0,u(x,t_0))} > 0$$

on $B \cap (\Omega \times \{t_0\})$. Also, $w(x_0, t_0) = 0$, $w(x, t_0) < 0$ on $B \cap (\Omega \times \{t_0\})$, and $\partial\Omega$ satisfies the interior sphere condition. By the Hopf Lemma, $\frac{\partial w(x_0,t_0)}{\partial \eta(x_0)} > 0$. However, by (4.15),

$$\frac{\partial w(x_0, t_0)}{\partial \eta(x_0)} = \frac{\partial u(x_0, t_0)}{\partial \eta(x_0)} \bullet N_0 = h(x_0, t_0, u_0) \bullet N_0 \leq 0,$$

a contradiction. It must be that Σ is positively invariant. □

If the strong flux condition is replaced by the weak flux condition, then the last theorem is no longer true. For example, consider

$$\begin{aligned} &u_t = \Delta u + (u-1)^{2/3}, \quad (x,t) \in \Pi_T, \quad T > 1 \\ &u(x, 0) = 0, \quad x \in \overline{\Omega} \\ &\tfrac{\partial u(x,t)}{\partial \eta(x)} = 0, \quad (x,t) \in \overline{S}_T. \end{aligned}$$

Let $\Sigma = (-1, 1)$; then $N(x, t, \pm 1) = \pm 1$ for any $(x,t) \in \overline{\Pi}_T$ and $f(x,t,u) = (u-1)^{2/3}$. At $u = 1 \in \partial\Sigma$, $N(x,t,1)f(x,t,1) = 0$, and at $u = -1 \in \partial\Sigma$, $N(x,t,-1)f(x,t,-1) = -2^{2/3} < 0$, so the weak flux condition is satisfied. However, a solution to the IBVP is $u(x,t) = 1 + (t-1)^3$. At $t = 1$, $u = 1$, but for $t \in (1, T]$, $u(x,t) > 1$ and Σ is not invariant.

4.3 Existence

The theorems in this section rely on compact operators associated with the initial-boundary value problems. We will construct these operators and then use Leray-Schauder degree theory to obtain solutions which lie in an invariant set.

For a function $v : \overline{\Pi}_T \to \mathbb{R}^m$, define $Fv : \overline{\Pi}_T \to \mathbb{R}^m$ by $Fv(x,t) = f(x,t,v(x,t))$. Let F denote the nonlinear operator which maps v to FV. Since f is Hölder continuous, $F : C^a(\overline{\Pi}_T) \to C^{a\alpha}(\overline{\Pi}_T)$ is a bounded mapping for any $a \in [0,1]$. In addition, $F : C(\overline{\Pi}_T) \to C(\overline{\Pi}_T)$ is continuous. By the boundedness of F we have $|Fv|_{C^{a\alpha}(\overline{\Pi}_T)} \leq \gamma |v|_{C^a(\overline{\Pi}_T)}$ for a generic constant γ.

Also, for the matrix $M(x,t,u)$, define Mv by $Mv(x,t) = M(x,t,v(x,t))$. Let M denote the nonlinear operator which maps v to Mv. The matrix $M(x,t,u)$ is also Hölder continuous, so the operator $M : C^a(\overline{\Pi}_T) \to C^{a\alpha}(\overline{\Pi}_T)$ is a bounded mapping for any $a \in [0,1]$ and is continuous for $a = 0$. Moreover, $|Mv|_{C^{a\alpha}(\overline{\Pi}_T)} \leq \gamma |v|_{C^a(\overline{\Pi}_T)}$ for a generic constant γ.

Define $B_a = \{v \in C^a(\overline{\Pi}_T) : v(x,0) = \phi(x) \text{ for } x \in \overline{\Omega}\}$ where $\phi(x)$ is the initial function in (4.9). For $v \in B_a$, consider the system

$$u_t = Mv(x,t)Eu(x,t) + Fv(x,t), \quad (x,t) \in \Pi_T \tag{4.16}$$
$$u(x,0) = \phi(x), \quad x \in \overline{\Omega} \tag{4.17}$$
$$u(x,t) = \psi(x,t), \quad (x,t) \in S_T \tag{4.18}$$

where we now assume that $M(x,t,u) = \text{diag}\{M_\ell(x,t,u) : \ell = 1,\ldots,m\}$ whose diagonal entries are positive. Consequently, we have an uncoupled system of linear equations with coefficients in $C^{a\alpha}(\overline{\Pi}_T)$ where each equation is uniformly parabolic. The standard linear theory implies that each equation in (4.16)-(4.17)-(4.18) has a solution $u_\ell \in C^{2+a\alpha}(\overline{\Pi}_T)$, $\ell = 1,\ldots,m$, which satisfies

$$|u_\ell|_{C^{2+a\alpha}(\overline{\Pi}_T)} \leq c(|Fv|_{C^{a\alpha}(\overline{\Pi}_T)} + |\phi|_{C^{2+a\alpha}(\overline{\Omega})} + |\psi|_{C^{2+a\alpha}(\overline{S}_T)}) \tag{4.19}$$

where c is some constant [LAD, Thm.5.2, pg.320].

Define the mapping $K : B_a \to C^{2+a\alpha}(\overline{\Pi}_T)$ by $Kv = u$; then K is a bounded operator for each $a \in (0,1]$ and

$$|Kv|_{C^{2+a\alpha}(\overline{\Pi}_T)} \leq c(\gamma |v|_{C^a(\overline{\Pi}_T)} + |\phi|_{C^{2+\alpha}(\overline{\Omega})} + |\psi|_{C^{2+\alpha}(\overline{S}_T)})$$

for some constant c.

Lemma 4.9 *The operator K can be extended to $K : C(\overline{\Pi}_T) \to C(\overline{\Pi}_T)$ and is continuous and compact.*

Proof. The operator K can be extended to $K : C(\overline{\Pi}_T) \to W_q^{2,1}(\Pi_T)$ for $q > 1$ sufficiently large, and is continuous. The extension follows from [LAD, Thm.9.1, pg.341]: For $v \in C(\overline{\Pi}_T)$, (4.16)-(4.17)-(4.18) has a unique solution $u = Kv \in W_q^{2,1}(\Pi_T)$ which satisfies the condition

$$|Kv|_{W_q^{2,1}(\Pi_T)} \leq c(|Fv|_{L_q(\Pi_T)} + |\phi|_{W_q^{2-2/q}(\Omega)} + |\psi|_{W_q^{2-1/q,1-1/(2q)}(S_T)}) \tag{4.20}$$

for some constant c. Using the estimates for F and M, and analyzing the proofs in [LAD, Ch.IV, Sec.9], one can replace the right-hand side of (4.20) by another bound:

$$|Kv|_{W_q^{2,1}(\Pi_T)} \leq c(\gamma|v|_{C(\overline{\Pi}_T)} + |\phi|_{C^{2+\alpha}(\overline{\Omega})} + |\psi|_{C^{2+\alpha}(\overline{S}_T)}). \tag{4.21}$$

Let $v_1, v_2 \in C(\overline{\Pi}_T)$ with $|v_1 - v_2|_{C(\overline{\Pi}_T)} < 1$. The functions Kv_1 and Kv_2 are solutions to (4.16)- (4.17)-(4.18), so

$$\begin{aligned}(Kv_1 - Kv_2)_t &= Mv_1E(Kv_1 - Kv_2) + Fv_1 - Fv_2 \\ &\quad +(Mv_1 - Mv_2)EKv_2,\end{aligned} \tag{4.22}$$

and $Kv_1 - Kv_2$ is a solution to (4.22) with zero initial and boundary conditions. Therefore,

$$\begin{aligned}&|Kv_1 - Kv_2|_{W_q^{2,1}(\Pi_T)} \\ &\leq c|Fv_1 - Fv_2 + (Mv_1 - Mv_2)EKv_2|_{L_q(\Pi_T)} \\ &\leq c(|Fv_1 - Fv_2|_{C(\overline{\Pi}_T)} + \gamma|Mv_1 - Mv_2|_{C(\overline{\Pi}_T)}|Kv_2|_{W_q^{2,1}(\Pi_T)})\end{aligned} \tag{4.23}$$

just as in the construction of (4.20) and (4.21). Since Kv_2 is a solution to (4.16), inequality (4.21) is valid with v replaced by v_2. Note that $|v_2|_{C(\overline{\Pi}_T)} \leq |v_1|_{C(\overline{\Pi}_T)} + 1$, so $|Kv_2|_{W_q^{2,1}(\Pi_T)} \leq \gamma$ for a generic constant depending on $|v_1|_{C(\overline{\Pi}_T)}$. Therefore (4.23) implies

$$|Kv_1 - Kv_2|_{W_q^{2,1}(\Pi_T)} \leq c(|Fv_1 - Fv_2|_{C(\overline{\Pi}_T)} + |Mv_1 - Mv_2|_{C(\overline{\Pi}_T)})$$

for some constant c, so $K : C(\overline{\Pi}_T) \to W_q^{2,1}(\Pi_T)$ is continuous.

From standard functional analysis [ADA], $W_q^{2,1}(\Pi_T)$ is compactly embedded in $C^{1+b}(\overline{\Pi}_T)$ for $0 < b < 1 - \frac{n+2}{q}$, and $C^{1+b}(\overline{\Pi}_T)$ is compactly embedded in $C(\overline{\Pi}_T)$, so extend K to $K : C(\overline{\Pi}_T) \to C(\overline{\Pi}_T)$, which is compact and continuous. □

Theorem 4.10 *Let $\Sigma \subset \mathbb{R}^m$ be an open bounded convex set which contains* 0. *Let the diagonal matrix M satisfy the eigenvalue condition (4.12), and let f satisfy the strong flux condition (4.13), with the same outer normal vector N. If $\phi \in C^{2+\alpha}(\overline{\Omega}, \Sigma)$ and $\psi \in C^{2+\alpha}(\overline{S}_T, \Sigma)$, then IBVP (4.8)-(4.9)-(4.10) has a solution $u \in C^{2+\alpha}(\overline{\Pi}_T, \Sigma)$.*

Proof. Let $K : C(\overline{\Pi}_T) \to C(\overline{\Pi}_T)$ be the completely continuous operator constructed in Lemma 4.9. If $u = \lambda Ku$ for some $\lambda \in (0, 1]$, then $u \in C^{2+\alpha}(\overline{\Pi}_T)$ is a solution of

$$u_t = MuEu + \lambda Fu, \quad (x,t) \in \Pi_T \tag{4.24}$$

$$u(x,0) = \lambda\phi(x), \quad x \in \overline{\Omega} \tag{4.25}$$

$$u(x,t) = \lambda\psi(x,t), \quad (x,t) \in \overline{S}_T. \tag{4.26}$$

The hypotheses of Theorem 4.7 are satisfied, so the solution u has the property $u(\overline{\Pi}_T) \subset \Sigma$.

We need only show existence to $u = \lambda K u$ for some $\lambda \in (0,1]$. Define $m(\Sigma) = \max\{|x| : x \in \overline{\Sigma}\}$ and $U = \{u \in C(\overline{\Pi}_T) : |u|_{C(\overline{\Pi}_T)} < m(\Sigma) + 1\} \neq \emptyset$. Consider the completely continuous operator $I - \lambda K$. For $\lambda \in [0,1]$, the Leray-Schauder degree, $\deg(I - \lambda K, U, 0)$, is defined. To see this, suppose there is some $u \in \partial U$ such that $(I - \lambda K)u = 0$ for some λ. If $\lambda = 0$, then $u = 0 \notin \partial U$. If $\lambda \neq 0$, then $u = \lambda K u$ is a solution to (4.24)- (4.25)-(4.26), so $u(\overline{\Pi}_T) \subset \Sigma$ and $|u|_{C(\overline{\Pi}_T)} \leq m(\Sigma)$; that is, $u \notin \partial U$. Thus, $(I - \lambda K)u \neq 0$ for all $u \in \partial U$ and $\deg(I - \lambda K, U, 0)$ is defined.

By homotopy invariance, $\deg(I - \lambda K, U, 0) = \deg(I, U, 0) = 1$ for $\lambda \in [0,1]$. Since $\deg(I - K, U, 0) \neq 0$, $u = Ku$ has at least one solution $u \in U$. Thus, $u \in C^{2+\alpha}(\overline{\Pi}_T)$ is a solution to (4.24)-(4.25)-(4.26) with $\lambda = 1$. □

The same result is true with the strong flux condition $N \bullet f < 0$ replaced by the weak flux condition $N \bullet f \leq 0$.

Theorem 4.11 *Let $\Sigma \subset \mathbb{R}^m$ be an open bounded convex set which contains 0. Let the diagonal matrix M satisfy the eigenvalue condition (4.12), and let f satisfy the weak flux condition, with the same outer normal vector N. If $\phi \in C^{2+\alpha}(\overline{\Omega}, \Sigma)$ and $\psi \in C^{2+\alpha}(\overline{S}_T, \Sigma)$, then IBVP (4.8)-(4.9)-(4.10) has a solution $u \in C^{2+\alpha}(\overline{\Pi}_T, \Sigma)$.*

Proof. For $\varepsilon \in (0, \frac{1}{2})$ consider

$$\begin{aligned} &u_t = (1-\varepsilon)(MuEu + Fu) - \varepsilon(u - \phi), \quad (x,t) \in \Pi_T \\ &u(x,0) = \phi(x), \quad x \in \overline{\Omega} \\ &u(x,t) = (1-\varepsilon)\psi(x,t) + \varepsilon\phi(x), \quad (x,t) \in S_T. \end{aligned} \tag{4.27}$$

Since $\phi(x) \in \Sigma$ and Σ is open, $(u(x,t) - \phi(x)) \bullet N(x,t,u) > 0$ for $u \in \partial\Sigma$ and $(x,t) \in \Pi_T$. Consequently,

$$N(x,t,u) \bullet [(1-\varepsilon)f(x,t,u) - \varepsilon(u - \phi(x))] < 0$$

and

$$N^T(x,t,u)(1-\varepsilon)M(x,t,u) = (1-\varepsilon)\lambda(x,t,u)N^T(x,t,u)$$

where $(1-\varepsilon)\lambda(x,t,u) > 0$. IBVP (4.27) satisfies the hypotheses of Theorems 4.7 and 4.10, so there is a solution $u_\varepsilon \in C^{2+\alpha}(\overline{\Pi}_T, \Sigma)$ for each $\varepsilon \in (0, \frac{1}{2})$.

As a solution in $W_q^{2,1}(\Pi_T)$, u_ε satisfies inequality (4.21). Since Σ is invariant, and since M and F are bounded continuous operators, (4.21) implies $|u_\varepsilon|_{W_q^{2,1}(\Pi_T)}$ is bounded independently of ε. Since $W_q^{2,1}(\Pi_T)$ is compactly embedded in $C^{1+b}(\overline{\Pi}_T)$ for some $b \in (0,1)$, $\{u_\varepsilon : 0 < \varepsilon < \frac{1}{2}\}$ is uniformly bounded in $C^{1+b}(\overline{\Pi}_T)$.

As a solution in $C^{2+\alpha}(\overline{\Pi}_T)$, u_ε satisfies inequality (4.19). The boundedness of $\{|u_\varepsilon|_{C^{1+b}(\overline{\Pi}_T)} : 0 < \varepsilon < \frac{1}{2}\}$ and the boundedness of M and

F imply that $\{|u_\varepsilon|_{C^{2+\alpha}(\overline{\Pi}_T)} : 0 < \varepsilon < \frac{1}{2}\}$ is uniformly bounded. Since $C^{2+\alpha}(\overline{\Pi}_T)$ is compactly embedded in $C^{2,1}(\overline{\Pi}_T)$, there is a subsequence $\{u_{\varepsilon_k}\}_{k=0}^{\infty}$ converging in norm to $u \in C^{2,1}(\overline{\Pi}_T)$ such that $\varepsilon_k \to 0$ as $k \to \infty$. The quantities $M(x,t,u_{\varepsilon_k}(x,t))$ and $f(x,t,u_{\varepsilon_k}(x,t))$ converge uniformly to $M(x,t,u(x,t))$ and $f(x,t,u(x,t))$, respectively, since M and f are continuous. Taking the limit as $k \to \infty$ in (4.27), with u replaced by ε_k, we see that $u \in C^{2,1}(\overline{\Pi}_T, \Sigma)$ is a solution to (4.8)-(4.9)-(4.10).

Since the coefficients M and Fu are in $C^{\alpha}(\overline{\Pi}_T)$, the standard linear theory shows that $u \in C^{2+\alpha}(\overline{\Pi}_T)$. □

Theorem 4.12 *Let $\Sigma \subset \mathbb{R}^m$ be a compact convex set containing* 0 *(with possibly empty interior relative to $\mathbb{R}^m$). Let the diagonal matrix M satisfy the eigenvalue condition (4.12) with eigenvalue λ, and let f satisfy the weak flux condition, with the same outer normal vector $N = N(u)$. If $\phi \in C^{2+\alpha}(\overline{\Omega}, \Sigma)$ and $\psi \in C^{2+\alpha}(\overline{S}_T, \Sigma)$, then IBVP (4.8)-(4.9)-(4.10) has a solution $u \in C^{2+\alpha}(\overline{\Pi}_T, \Sigma)$.*

Proof. Let P be the projection operator which maps each point $u \in \mathbb{R}^m$ to the nearest point $Pu \in \Sigma$; then $|Pu - u| = \inf\{|s - u| : s \in \Sigma\}$ and P is uniformly Lipschitz continuous.

For each $q \in (0,1]$, define $\Sigma_q = \{u \in \mathbb{R}^m : \text{dist}(u, \Sigma) < q\} \supset \Sigma$, where Σ_q is open, bounded, and convex. If $u \in \partial\Sigma_q$, then $Pu \in \partial\Sigma$ and $u - Pu$ is an outer normal to $\partial\Sigma_q$ at u and to $\partial\Sigma$ at Pu. For $q \in (0,1)$ consider

$$\begin{aligned} u_t &= MPuEu + FPu, \quad (x,t) \in \Pi_T \\ u(x,0) &= \phi(x) \in C^{2+\alpha}(\overline{\Omega}, \Sigma_q) \\ u(x,t) &= \psi(x,t) \in C^{2+\alpha}(\overline{S}_T, \Sigma_q). \end{aligned} \tag{4.28}$$

For $u \in \partial\Sigma_q$, we have $N(u) = u - Pu$ and $\lambda(x,t,Pu) > 0$. The hypotheses of Theorem 4.11 are satisfied, so (4.28) has a solution $u_q \in C^{2+\alpha}(\overline{\Pi}_T, \Sigma_q)$.

Since the sets Σ_q are nested, the set $\{u_q : 0 < q < 1\}$ is uniformly bounded in $C(\overline{\Pi}_T)$ by $|u_q|_{C(\overline{\Pi}_T)} \leq \sup\{|u| : u \in \Sigma_1\}$. Similar to the proof of Theorem 4.11, $\{u_q : 0 < q < 1\}$ is uniformly bounded in $C^{2+\alpha}(\overline{\Pi}_T)$, and there is a subsequence u_{q_k} which converges to a function $u \in C^{2,1}(\overline{\Pi}_T)$ with $q_k \to 0$ as $k \to \infty$. This function u is a solution to (4.8)-(4.9)-(4.10) and satisfies $u \in C^{2+\alpha}(\overline{\Pi}_T, \Sigma)$. □

The following existence results for IBVP (4.8)-(4.9)-(4.11) are given without proofs. The proofs are similar to those for IBVP (4.8)- (4.9)-(4.10) in that they use Leray-Schauder degree theory arguments applied to certain compact operators. However, the boundary condition (4.11) creates more technical difficulties. The proofs can be found in [TAL1].

Theorem 4.13 *Let $\Sigma \subset \mathbb{R}^m$ be an open bounded convex set which contains* 0*. Let M be a diagonal matrix satisfying the eigenvalue condition*

(4.12). Let f satisfy the weak flux condition for IBVP (4.8)-(4.9)-(4.11) with the same outer normal vector N, and where $h(x,t,u) = b(x)u$ and $b \in C^{1+\alpha}(\partial\Omega, (-\infty, 0])$.

If f satisfies the growth condition

$$|f(x,t,u,p)| \leq c(1 + |p|^{2-\varepsilon})$$

for $(x,t,u) \in \overline{\Pi}_T \times \Sigma$, c constant, and $\varepsilon \in (0,1)$, and if $\phi \in C^{2+\alpha}(\overline{\Omega}, \Sigma)$ with $\frac{\partial\phi(x)}{\partial\eta(x)} = b(x)\phi(x)$ for $x \in \partial\Omega$, then IBVP (4.8)-(4.9)-(4.11) has a solution $u \in C^{2+\alpha}(\overline{\Pi}_T, \Sigma)$.

Theorem 4.14 *Let $\Sigma \subset \mathbb{R}^m$ be a compact convex set which contains 0. Let M be a diagonal matrix satisfying the eigenvalue condition (4.12). Let f satisfy the weak flux condition for IBVP (4.8)-(4.9)- (4.11) with the same outer normal vector N, and where $h(x,t,u) \equiv 0$.*

If f satisfies the growth condition

$$|f(x,t,u,p)| \leq c(1 + |p|^{2-\varepsilon})$$

for $(x,t,u) \in \overline{\Pi}_T \times \Sigma$, c constant, and $\varepsilon \in (0,1)$, and if $\phi \in C^{2+\alpha}(\overline{\Omega}, \Sigma)$ with $\frac{\partial\phi(x)}{\partial\eta(x)} = 0$ for $x \in \partial\Omega$, then IBVP (4.8)-(4.9)-(4.11) has a solution $u \in C^{2+\alpha}(\overline{\Pi}_T, \Sigma)$.

4.4 Applications

The invariance and existence results developed in Sections 4.2 and 4.3 can be extended to problems with mixed boundary conditions. We give a brief outline of how this extension can be carried out.

For $v : \overline{\Pi}_T \to \mathbb{R}$, define the differential operator L_k by

$$L_k v = \frac{\partial v}{\partial t} - \left(\sum_{i,j} a_{kij}(x,t) \frac{\partial^2 v}{\partial x_i x_j} + \sum_i b_{ki}(x,t) \frac{\partial v}{\partial x_i} + c_k(x,t) v \right)$$

where $a_{kij}, b_{ki}, c_k \in C^{\alpha,\alpha/2}(\overline{\Pi}_T, \mathbb{R})$ for some $\alpha \in (0,1)$, and where $c_k \leq 0$. Assume that each L_k is uniformly parabolic. Define the operator L by $Lu = (L_1 u_1, \ldots, L_m u_m)$ for $u : \overline{\Pi}_T \to \mathbb{R}^m$.

As in the previous sections, let $f(x,t,u,p) \in C^{\alpha,\alpha/2,\alpha,\alpha}(\overline{\Pi}_T \times \mathbb{R}^m \times \mathbb{R}^{nm}, \mathbb{R}^m)$ where p is an $n \times m$ matrix. We consider the system of mixed initial-boundary value problems of the form

$$Lu = f(x,t,u,Du), \quad (x,t) \in \Pi_T \tag{4.29}$$

$$Bu = 0, \quad (x,t) \in \Gamma_T \tag{4.30}$$

where $Du = [\partial u_j/\partial x_i]_{n\times m}$ and $\Gamma_T = (\overline{\Omega}\times\{0\})\cup S_T$. The initial-boundary value operator B is defined by $Bu = (B_1u_1,\ldots,B_mu_m)$ with

$$B_iv(x,t) = \left\{\begin{array}{ll} v(x,0)-\phi_i(x), & x\in\overline{\Omega} \\ \frac{\partial v(x,t)}{\partial\eta(x)} - b_i(x,t)v(x,t), & (x,t)\in\overline{S}_T \end{array}\right\}$$

or

$$B_iv(x,t) = v(x,t)-\psi_i(x,t),\ \ (x,t)\in\Gamma_T,$$

where

$$\phi\in C^{2+\alpha}(\overline{\Omega},\mathrm{I\!R}^m),\ b\in C^{1+\alpha}(\overline{\Pi}_T,(-\infty,0]^m),$$

and

$$\eta(x)\in C^{1+\alpha}(\partial\Omega,\mathrm{I\!R}^m)$$

is an outer unit normal vector to $\partial\Omega$ at x.

Let $\alpha,\beta\in C^{2,1}(\overline{\Pi}_T,\mathrm{I\!R}^m)$ satisfy $\alpha(x,t)<\beta(x,t)$ for $(x,t)\in\overline{\Pi}_T$. Define $(\alpha,\beta)=\{u\in\mathrm{I\!R}^m : \alpha(x,t)<u<\beta(x,t) \text{ for } (x,t)\in\overline{\Pi}_T\}$. Similarly define $(\alpha,\beta]$, $[\alpha,\beta)$, and $[\alpha,\beta]$.

Let $v,w\in\mathrm{I\!R}^m$. Recall the definition for the replacement vector

$$\hat{v}^j(w) = (v_1,\ldots,v_{j-1},w_j,v_{j+1},\ldots,v_m).$$

The next two results can be found in [BEB3] and are used for two applications given later in this section.

Theorem 4.15 *Assume that*

$$L_k\alpha_k - f_k(x,t,\hat{u}^k(\alpha),D\hat{u}^k(\alpha)) < 0 < L_k\beta_k - f_k(x,t,\hat{u}^k(\beta),D\hat{u}^k(\beta))$$

for all $(x,t)\in\Pi_T$ *and* $\alpha_j\le u_j\le\beta_j$, $j\ne k$, $k=1,\ldots,m$; *then* $\Sigma=(\alpha,\beta)$ *is positively invariant relative to (4.29)-(4.30).*

Proof. The proof is essentially that of Theorems 4.7 and 4.8 with a few modifications. If the solution u to (4.29)-(4.30) satisfies $u(\overline{\Pi}_T)\not\subset\Sigma$, then there is an index k, a first time t_0, and a value $x_0\in\overline{\Omega}$ such that $u_k(x_0,t_0)=\beta_k(x_0,t_0)$ or $u_k(x_0,t_0)=\alpha_k(x_0,t_0)$. Without loss of generality assume that the latter case happens. Define $w=u_k-\alpha_k$. If the boundary condition B_k is Dirichlet, then the proof is similar to that of Theorem 4.7. If the boundary condition B_k is Robin, then the proof is similar to that of Theorem 4.8. □

Theorem 4.16 *Let* $\alpha,\beta\in C^{2,1}(\overline{\Pi}_T,\mathrm{I\!R}^m)$ *with* $\alpha(x,t)<\beta(x,t)$ *for all* $(x,t)\in\overline{\Pi}_T$. *Assume that*

$$L_k\alpha_k - f_k(x,t,\hat{u}^k(\alpha),D\hat{u}^k(\alpha)) \le 0 \le L_k\beta_k - f_k(x,t,\hat{u}^k(\beta),D\hat{u}^k(\beta))$$

for all $(x,t)\in\Pi_T$ *and* $\alpha_j\le u_j\le\beta_j$, $j\ne k$, $k=1,\ldots,m$. *If there is a continuous nondecreasing function* $\Phi:[0,\infty)\to(0,\infty)$ *such that* $s^2/\Phi(s)\to\infty$

as $s \to \infty$, and if f satisfies the growth condition $|f(x,t,u,p)| \le \Phi(|p|)$ for $(x,t,u) \in \overline{\Pi}_T \times [\alpha,\beta]$, and if all initial data have ranges in $[\alpha,\beta]$, then IBVP (4.29)-(4.30) has a solution $u \in C^{2,1}(\overline{\Pi}_T, [\alpha,\beta])$.

As an immediate consequnce, we have Theorem 3.1 for the scalar problem.

The first application we wish to consider is to a system which models certain chemical processes in which a gas is absorbed by a liquid which contains a substance that reacts with the dissolved gas [KAH]:

$$\begin{aligned}
&u_t = a\Delta u - cuv, \ v_t = b\Delta v - duv, \quad (x,t) \in \Pi_T \\
&u(x,0) = \phi_1(x), \ v(x,0) = \phi_2(x), \quad x \in \overline{\Omega} \\
&u(x,t) = \psi_1(x,t), \ v(x,t) = \psi_2(x,t), \quad (x,t) \in \overline{S}_T
\end{aligned}$$

where a, b, c, and d are positive, and ϕ_1, ϕ_2, ψ_1, and ψ_2 are nonnegative functions.

This system is a special case of a system

$$L_1 u = f_1(x,t,u,v), \quad L_2 v = f_2(x,t,u,v),$$

where

$$f_1(x,t,0,v) \ge 0, \quad f_2(x,t,u,0) \ge 0,$$

and where

$$f_1(x,t,M_1,v) \le 0, \quad f_2(x,t,u,M_2) \le 0$$

for some positive constants M_1 and M_2. By Theorem 4.16 there is a solution $(u,v) \in [\overline{0}, \overline{M}]$ where $\overline{0} = (0,0)$ and $\overline{M} = (M_1, M_2)$. For the special case above, choose

$$M_i = \max\{\sup_{x\in\Omega} \phi_i(x), \quad \sup_{(x,t)\in S_T} \psi_i(x,t)\}.$$

The second application is to the complete model for solid fuel (1.24)-(1.25):

$$\begin{aligned}
&\left.\begin{aligned}
&T_t - \Delta T = \varepsilon\delta y^m \exp\left(\tfrac{T-1}{\varepsilon T}\right) \\
&y_t - \beta\Delta y = -\varepsilon\delta\Gamma y^m \exp\left(\tfrac{T-1}{\varepsilon T}\right)
\end{aligned}\right\}, \quad (x,t) \in \Pi_T \\
&T(x,0) = 1, \ y(x,0) = 1, \quad x \in \overline{\Omega} \\
&T(x,t) = 1, \ \tfrac{\partial y(x,t)}{\partial \eta(x)} = 0, \quad (x,t) \in S_T
\end{aligned}$$

where ε, δ, β, and Γ are positive constants.

Let $\psi(x)$ be the solution to $-\Delta\psi = 1$ for $x \in \Omega$ and $\psi(x) = 1$ for $x \in \partial\Omega$. By the maximum principle, $\psi(x) > 0$ on $\overline{\Omega}$. Choose N such that $N > \varepsilon e^{1/\varepsilon}\delta$ and $N\psi(x) \ge \varepsilon$. Define $\xi(x,t) = (0,0)$ and $\rho(x,t) = (N\psi(x), 1)$, and define

$f_1(x,t,T,y) = \varepsilon\delta y^m \exp(\frac{T-1}{\varepsilon T})$ and $f_2(x,t,T,u) = -\varepsilon\delta\Gamma y^m \exp(\frac{T-1}{\varepsilon T})$. We have

$$\begin{aligned}
&\tfrac{\partial \xi_1}{\partial t} - \Delta\xi_1 - f_1(x,t,\xi_1,y) = 0\\
&\tfrac{\partial \xi_2}{\partial t} - \Delta\xi_2 - f_2(x,t,T,\xi_2) = 0\\
&\tfrac{\partial \rho_1}{\partial t} - \Delta\rho_1 - f_1(x,t,\rho_1,y)\\
&\quad = -N\Delta\psi - \varepsilon e^{1/\varepsilon}\delta y^m \exp(-\tfrac{1}{\varepsilon\rho_1}) \geq N - \varepsilon e^{1/\varepsilon}\delta > 0\\
&\tfrac{\partial \rho_2}{\partial t} - \Delta\rho_2 - f_1(x,t,T,\rho_2) = \varepsilon\delta\Gamma y^m \exp(\tfrac{T-1}{\varepsilon T}) > 0.
\end{aligned}$$

By Theorem 4.16, there exists a solution $(T,y) \in [\xi,\rho]$ for all $(x,t) \in \overline{\Omega} \times [0,\infty)$.

4.5 Comments

The comparison techniques in Theorem 4.1 and Corollary 4.2 can be traced back to Nagumo [NAG2],[NAG4]. The ideas were rediscovered by Westphal [WST]. The extension to parabolic systems with quasimonotone nonlinearities (Theorem 4.3) is due to Mlak [MLA1]. The use of upper and lower quasimonotone bounds on the nonlinearity to provide the comparison theorem 4.4 can be found in Chandra and Davis [CHA] and Fife [FIF].

Weinberger [WEI] proved that a closed convex set is invariant for the Dirichlet problem if the nonlinearity f satisfies the weak tangent condition. Chueh, Conley, and Smoller [CHU] extended the result to include sets which are Cartesian products of convex sets and the components of the operator L are the same for any set in the product. Moreover, the convexity condition is optimal. Bebernes and Schmitt [BEB1] generalized the invariance results to include nonlinearities which have gradient dependence.

Redheffer and Walter [RED] consider problems with more general domains and various boundary conditions, but f must satisfy a dissipative condition near the boundary of the set.

Amann [AMA3] considered gradient dependent nonlinearities and discussed invariance for systems as evolution equations in a Banach space. The methods used include semigroup theory. In this paper, the existence of a unique solution is obtained. The existence theorems in this chapter allow for nonuniqueness.

Generalizations of the results in this chapter appear in [YAN]. Apparently for the nonlinear boundary conditions, the condition of almost quadratic growth in the gradient component of f is the best that the ideas of the proof can handle.

5
Gaseous Ignition Models

We discuss in this chapter initial-boundary value problems of the form

$$u_t - a\Delta u = f(u) + g(t), \quad a \geq 0, \quad (x,t) \in \Omega \times (0,T)$$

with $u(x,0) = \phi(x)$ for $x \in \Omega$ and $u(x,t) = 0$ for $(x,t) \in \partial\Omega \times (0,T)$. The reactive-diffusive gaseous model (1.39)- (1.40) and the nondiffusive model (1.41)-(1.40) are special cases.

In Section 5.1, we begin by considering the case when $a > 0$ and g has a special functional dependence of the form $K \int_\Omega u_t(x,t)\,dx$. Such a dependency complicates the required analysis as it can be considered in an equivalent formulation (5.4) as a perturbation of both the diffusion and the reaction terms. The problem is cast in an abstract setting in Sections 5.2 and 5.3. Using a semigroup analysis, existence of a unique nonextendable solution is proved.

For $\Omega = B_1 \subset \mathbb{R}^n$, additional comparison results are obtained in Section 5.4. For zero initial data and $\delta > \delta_{FK}$, blowup occurs for (1.39)-(1.40) at a time $\sigma < T$ where T is the blowup time for the solid fuel ignition model.

In Section 5.5, the location of the blowup in $\Omega = B_R$ is discussed. Depending on the nonlinearity f, blowup can occur everywhere or at a single point.

In Section 5.6, the nondiffusive model ($a = 0$) is discussed. A very precise description of when and where blowup occurs is given.

5.1 The Reactive-Diffusive Ignition Model

Consider the partial differential equation

$$\theta_t - \Delta\theta = f(\theta) + \frac{\gamma - 1}{\gamma}\frac{1}{\mathrm{vol}(\Omega)}\int_\Omega \theta_t(y,t)\,dy, \quad (x,t) \in \Omega \times (0,\infty) \tag{5.1}$$

with initial-boundary conditions

$$\begin{aligned} &\theta(x,0) = \theta_0(x), \quad x \in \overline{\Omega} \\ &\theta(x,t) = 0, \quad (x,t) \in \partial\Omega \times (0,\infty) \end{aligned} \tag{5.2}$$

where Ω is a bounded domain in $\mathbb{R}^n$, $\theta_0(x)$ will be specified as needed, $\delta > 0$, $\gamma > 1$, and f satisfies $f(u) > 0$, $f'(u) \geq 0$, and $f''(u) \geq 0$.

If $\theta(x,t)$ is a solution of (5.1)-(5.2), then integrating (5.1) over Ω gives

$$\frac{1}{\gamma}\int_\Omega \theta_t(y,t)\,dy = \int_\Omega (\Delta\theta(y,t) + f(\theta(y,t))\,dy.$$

Consequently, (5.1) is equivalent to

$$\theta_t - \Delta\theta = f(\theta) + \frac{\gamma-1}{\mathrm{vol}(\Omega)}\int_\Omega (\Delta\theta(y,t) + f(\theta(y,t))\,dy. \tag{5.3}$$

By applying Green's identity to $\int_\Omega \Delta\theta\,dy$ in (5.3), we see that (5.1) is also equivalent to

$$\theta_t - \left(\Delta\theta + \frac{\gamma-1}{\mathrm{vol}(\Omega)}\int_\Omega \frac{\partial\theta}{\partial\nu}\,d\sigma\right) = f(\theta) + \frac{\gamma-1}{\mathrm{vol}(\Omega)}\int_\Omega f(\theta(y,t)\,dy \tag{5.4}$$

where $\nu(x)$ is the outer unit normal to $\partial\Omega$ at x, and $d\sigma$ is the element of surface area on $\partial\Omega$.

5.2 The Abstract Linear Problem

Consider the associated linear problem to (5.3)-(5.2) given by

$$\begin{aligned} &\theta_t = \Delta\theta + \tfrac{1}{c}\textstyle\int_\Omega \Delta\theta\,dy, \quad (x,t)\in\Omega\times(0,\infty)\\ &\theta(x,0) = \theta_0(x), \quad x\in\overline{\Omega}\\ &\theta(x,t) = 0, \quad (x,t)\in\partial\Omega\times(0,\infty)\end{aligned} \tag{5.5}$$

where $c = \frac{\mathrm{vol}(\Omega)}{\gamma-1} > 0$. We will prove that the right- hand side of (5.5a) is the infinitesimal generator of an analytic equicontinuous semigroup T.

Let $F = L^2(\Omega)$ and $E = F\oplus\mathbb{R}$. For $(f,\eta),(g,\xi)\in E$, define an inner product $\langle\cdot,\cdot\rangle$ on E by

$$\langle (f,\eta),(g,\xi)\rangle = \int_\Omega fg\,dx + c\eta\xi.$$

Thus, E is a Hilbert space with norm $\|(f,\eta)\| = \sqrt{\langle (f,\eta),(f,\eta)\rangle}$. We first consider a related problem

$$\begin{aligned} &f_t = \Delta f, \quad (x,t)\in\Omega\times(0,\infty)\\ &\eta_t = -\tfrac{1}{c}\textstyle\int_{\partial\Omega}\frac{\partial f}{\partial\nu}\,d\sigma, \quad t\in(0,\infty)\\ &f(x,t) = \eta(t), \quad (x,t)\in\partial\Omega\times(0,\infty).\end{aligned} \tag{5.6}$$

This can be expressed as

$$\frac{d}{dt}(f,\eta) = \hat{A}(f,\eta) \tag{5.7}$$

where $\hat{A}(f,\eta) = (\Delta f, -\frac{1}{c}\int_{\partial\Omega} \frac{\partial f}{\partial \nu}\, d\sigma)$ and where condition (5.6c) is assumed to hold.

Theorem 5.1 *The function $\hat{A}$ generates an analytic contraction semigroup $S := \{S(t) : t \geq 0\}$ on E.*

Proof. It is apparent that $\hat{A}$ is linear, closed, and densely defined on E. We also claim that $\hat{A}$ is self-adjoint and dissipative.

For $(f,\eta),(g,\xi) \in \mathrm{Dom}(\hat{A})$, we have the following computation.

$$\begin{aligned}
\langle \hat{A}(f,\eta),(g,\xi)\rangle &= \int_\Omega g\Delta f\, dx + c(-\tfrac{1}{c}\int_{\partial\Omega} f_\nu\, d\sigma)\xi \\
&= -\int_\Omega \nabla f \bullet \nabla g\, dx + \int_{\partial\Omega} f_\nu g\, d\sigma - \int_{\partial\Omega} f_\nu \xi\, d\sigma \\
&= -\int_\Omega \nabla f \bullet \nabla g\, dx \\
&= \langle (f,\eta), \hat{A}(g,\xi)\rangle
\end{aligned}$$

where Green's identity and the fact that $g(x,t) = \xi(t)$ on $\partial\Omega \times (0,\infty)$ have been used. This proves self-adjointness.

In particular, from the above we have that

$$\langle \hat{A}(f,\eta),(f,\eta)\rangle = -\int_\Omega |\nabla f|^2 dx \leq 0.$$

By the Lumer-Phillips Theorem [YOS, pp.250-251], we conclude that $\hat{A}$ generates a contraction semigroup S with $\|S(t)\| \leq 1$ for all $t \geq 0$. But the self-adjointness of $\hat{A}$ implies that its complex extension $\tilde{A}$ is Hermitian. Hence, its numerical range $\{\langle \hat{A}(f,\eta),(f,\eta)\rangle\}$ lies on the negative real axis. By [MAR, Prop.3.2, pg.293], this implies that S is analytic. □

We now relate the generator $\hat{A}$ and its semigroup S to the original problem (5.5). Consider the canonical injection $\imath : F \to E$ defined by $\imath(f) = (f,0)$. Define the projection $\pi : E \to F$ by $\pi(f,\eta) = f - \tilde{\eta}$ where $\tilde{\eta}(x) = \eta$ for all $x \in \Omega$. Clearly π is a continuous projection of E onto F. For $t \geq 0$, define the operator $T(t) : F \to F$ by

$$T(t)f = \pi \circ S(t) \circ \imath(f). \tag{5.8}$$

Theorem 5.2 *The set $T = \{T(t) : t \geq 0\}$ is an analytic semigroup on F with generator A given by*

$$Af = \Delta f + \frac{1}{c}\int_\Omega \Delta f\, dx. \tag{5.9}$$

Proof. The linearity and continuity of each $T(t)$ are obvious. The mapping $t \to T(t)$ is a composition of two linear and continuous mappings with an analytic mapping. Therefore, it is analytic (and hence uniformly continuous).

To show T is a semigroup, consider the following. For any $\theta \in F$ we have $T(0)\theta = \pi \circ S(0) \circ \imath(\theta)$. If $t, s \geq 0$, then

$$\begin{aligned} T(t) \circ T(s)\theta &= \pi \circ S(t) \circ \imath \circ \pi \circ S(s) \circ \imath(\theta) \\ &= \pi \circ S(t) \circ S(s) \circ \imath(\theta) \\ &\quad + \pi \circ S(t) \circ [\imath \circ \pi \circ S(s) \circ \imath(\theta) - S(s) \circ \imath(\theta)]. \end{aligned}$$

Since $\ker(\pi) = \{(f, \eta) : f(x) = \eta,\ x \in \Omega,\ \eta \in \mathbb{R}\}$, we have for $(f, \eta) \in E$

$$\begin{aligned} (f, \eta) - \imath \circ \pi(f, \eta) &= (f, \eta) - \imath(f - \tilde{\eta}) \\ &= (f, \eta) - (f - \tilde{\eta}, 0) \\ &= (\tilde{\eta}, \eta) \\ &\in \ker(\pi). \end{aligned}$$

But the kernel is invariant under $S(t)$, so

$$\pi \circ S(t) \circ [\imath \circ \pi \circ S(s) \circ \imath(\theta) - S(s) \circ \imath(\theta)] = 0,$$

and thus,

$$\begin{aligned} T(t) \circ T(s)\theta &= \pi \circ S(t) \circ S(s) \circ \imath(\theta) \\ &= \pi \circ S(t + s) \circ \imath(\theta) \\ &= T(t + s)\theta. \end{aligned}$$

That is, T is a semigroup on F.

We now prove that A is the generator of T. Assume that $\theta \in C^2(\Omega)$ and $\theta(x) = 0$ for $x \in \partial\Omega$; then $\imath(\theta) = (\theta, 0) \in \mathrm{Dom}(\hat{A})$ and $\pi \circ \imath(\theta) = \theta$. Thus,

$$\begin{aligned} \lim_{h \to 0} \tfrac{T(h)\theta - \theta}{h} &= \lim_{h \to 0} \tfrac{\pi \circ S(h) \circ \imath(\theta) - \theta}{h} \\ &= \lim_{h \to 0} \tfrac{\pi \circ [S(h) \circ \imath(\theta) - \imath(\theta)]}{h} \\ &= \pi \circ \left(\lim_{h \to 0} \tfrac{S(h) \circ \imath(\theta) - \imath(\theta)}{h} \right) \\ &= \pi \circ \hat{A} \circ \imath(\theta) \\ &= \pi \circ \hat{A}(\theta, 0) \\ &= \pi \left(\Delta\theta, -\tfrac{1}{c} \int_{\partial\Omega} \theta_\nu \, d\sigma \right) \\ &= \Delta\theta + \tfrac{1}{c} \int_\Omega \Delta\theta \, dx \\ &= A(\theta) \end{aligned}$$

where we again have used Green's identity. □

We now show that certain subsets of F (respectively E) are invariant relative to the semigroup $T(s)$ (respectively S) generated by A (respectively $\hat{A}$). We define a set $D \subset E$ (respectively F) to be invariant for S (respectively T) if $S(t)$ (respectively $T(t)$) maps D into itself for all $t \geq 0$.

Lemma 5.3 *For any $b \in \mathbb{R}$, the set*

$$\Lambda_b := \{(f, \eta) \in E : f(x) \leq b \ \text{for} \ x \in \Omega, \ \eta \leq b\}$$

is invariant under S.

Proof. The set Λ_b is clearly a closed convex subset of E. By the result [MAR, Prop.5.3, pg.304], it suffices to show that if $\lambda > 0$ and $(f, \eta) \in \Lambda_b$, then $(g, \xi) = (I - \lambda \hat{A})^{-1}(f, \eta) \in \Lambda_b$.

Assume that $(g, \xi) \in \text{Dom}(\hat{A})$ solves $f(x) = g(x) - \lambda \Delta g(x)$ for $x \in \Omega$ and $\eta = \xi + \frac{\lambda}{c} \int_{\partial\Omega} g_\nu \, d\sigma$ with $g(x) = \xi$ for $x \in \partial\Omega$. Let $g(\bar{x}) = \sup\{g(x) : x \in \Omega\}$. Two cases must be considered. First, if $\bar{x} \in \Omega$, then $\Delta g(\bar{x}) \leq 0$ and so $g(\bar{x}) \leq g(\bar{x}) - \lambda \Delta g(\bar{x}) = f(\bar{x}) \leq b$. Second, if $\bar{x} \in \partial\Omega$, then $g(\bar{x}) = \xi$ and $\int_{\partial\Omega} g_\nu \, d\sigma \geq 0$ since g assumes its maximum on $\partial\Omega$. This yields the condition $g(\bar{x}) = \xi \leq \xi + \frac{\lambda}{c} \int_{\partial\Omega} g_n \, d\sigma = \eta \leq b$. □

Corollary 5.4 *For any $a, b \in \mathbb{R}$ such that $a < b$, the set*

$$\Lambda_{a,b} := \{(f, \eta) \in E : a \leq f(x) \leq b \ \text{for} \ x \in \Omega, \ a \leq \eta \leq b\}$$

is invariant under S.

We can now prove an invariance result for the semigroup T. Let $\| \cdot \|_\infty$ denote the essential sup norm on $L^\infty(\Omega)$ and let f_+ and f_- be the positive and negative parts of $f \in L^\infty(\Omega)$.

Theorem 5.5 *For $\rho > 0$, the set*

$$D_\rho := \{f \in F : \|f_+\|_\infty + \|f_-\|_\infty \leq \rho\}$$

is invariant under T.

Proof. Let $a = \|f_-\|_\infty$ and $b = \|f_+\|_\infty$ where $a + b \leq \rho$; then $(f, 0) \in \Lambda_{-a,b}$ and by Corollary 5.4, for all $t \geq 0$ we have $(g, \xi) = S(t)(f, 0) \in \Lambda_{-a,b}$. So

$$\|[T(t)f]_+\|_\infty = \|(g - \tilde{\xi})_+\|_\infty \leq b - \xi$$

and

$$\|[T(t)f]_-\|_\infty = \|(g - \tilde{\xi})_-\|_\infty \leq a + \xi$$

where $\tilde{\xi}(x) = \xi$ for all $x \in \Omega$. Thus, $T(t)f \in D_\rho$ for all $t \geq 0$. □

Corollary 5.6 *If $f, g \in F = L^2(\Omega)$ and $\sup\{|f(x) - g(x)| : x \in \Omega\} < \varepsilon$, then $\sup\{|T(t)[f(x) - g(x)]| : x \in \Omega\} \leq 2\varepsilon$ for all $t \geq 0$.*

We now consider the case where $\Omega = B_1 \subset \mathbb{R}^n$. Let $F_s = \{v \in F :$ v is radially symmetric$\}$. For each $v \in \text{Dom}(A) \cap F_s$, we have

$$\Delta v(x) = v''(r) + \frac{n-1}{r} v'(r), \quad v'(0) = v(1) = 0, \quad \text{and} \quad \int_{\partial\Omega} v_\nu \, d\sigma = w_n v'(1)$$

where $r = |x|$ and w_n is the surface area of the unit ball in $\mathbb{R}^n$. Let

$$\Gamma := \{v \in F_s : v(r_1) \geq v(r_2) \text{ for } 0 \leq r_1 \leq r_2 \leq 1, \; v(1) = 0\}$$

be the set of nonnegative, nondecreasing, radially symmetric functions with domain $[0, 1]$.

Theorem 5.7 *The set Γ is invariant under T.*

Proof. The set Γ is a closed convex cone in F. By [MAR, Prop.5.3, pg.304], it suffices to show that $v = (I - \lambda A)^{-1}\theta \in \Gamma$ if $\lambda > 0$ and $\theta \in \Gamma$.

Assume for some $0 \leq r_1 < r_2 \leq 1$ that $v(r_1) < v(r_2)$; then v attains a local minimum $v(\bar{r})$ at some point $\bar{r} \in [0, r_2)$. Two cases are possible.

If v is increasing on $[\bar{r}, 1]$, then $v(\bar{r}) < 0$ since $v(1) = 0$, $v'(\bar{r}) = 0$, $v''(\bar{r}) \geq 0$, and $v'(1) \geq 0$. Hence,

$$\theta(\bar{r}) = v(\bar{r}) - \lambda[v''(\bar{r}) + \frac{w_n}{c} v'(1)] < 0$$

which contradicts $\theta \in \Gamma$.

If v is not increasing on $[\bar{r}, 1]$, then v has a local maximum $v(\tilde{r}) > v(\bar{r})$ at some point $\tilde{r} \in (\bar{r}, 1)$. This implies that $v'(\tilde{r}) = v'(\bar{r}) = 0$, $v''(\bar{r}) \geq 0$, and $v''(\tilde{r}) \leq 0$. Hence,

$$\theta(\tilde{r}) - \theta(\bar{r}) = v(\tilde{r}) - v(\bar{r}) - \lambda[v''(\tilde{r}) - v''(\bar{r})] > 0$$

which again contradicts $\theta \in \Gamma$. We conclude that $v(r_1) \geq v(r_2)$ for $0 \leq r_1 < r_2 \leq 1$, so $v \in \Gamma$ and Γ is invariant under T. □

Corollary 5.8 *If $f, g \in F_s$ have the properties $f \geq g$ and $f - g \in \Gamma$, then $T(t)f \geq T(t)g$ for all $t \geq 0$.*

Proof. Since Γ is invariant under T by Theorem 5.7, $T(t)(f-g)(x) \in \Gamma$ and so $T(t)(f - g) \geq 0$. □

Let $A_\Delta = \Delta$ be the n-dimensional Laplacian operator with the same domain as A. Let T_Δ be the semigroup generated by A_Δ.

Corollary 5.9 *Let $\theta_0 \in \Gamma$; then $T_\Delta(t)\theta_0 \geq T(t)\theta_0$ for all $t \geq 0$.*

Proof. By Theorem 5.7, $T(t)\theta_0 \geq 0$ for all $t \geq 0$. Hence,

$$\frac{1}{c} \int_\Omega \Delta\theta_0 \, dx = \frac{1}{c} \int_\Omega \frac{\partial \theta_0}{\partial \nu} \, d\sigma \leq 0$$

and $T(t)\theta_0$ is a lower solution to $\frac{d\theta}{dt} = A_\Delta \theta$. □

5.3 The Abstract Perturbed Problem

Let us now consider the perturbed problem

$$\frac{d\theta}{dt} = A\theta + B(\theta), \quad \theta(0) = \theta_0 \tag{5.10}$$

where

$$B(\theta) = f(\theta) + \frac{\gamma - 1}{\text{vol}(\Omega)} \int_\Omega f(\theta(y))\, dy$$

with $f(u) > 0$, $f'(u) \geq 0$, and $f''(u) \geq 0$. We first prove existence locally of a classical solution. To do this we use the following lemma.

Lemma 5.10 *Consider the system of differential inequalities*

$$\begin{aligned} u' &\leq \sqrt{\gamma V}\, f(u + \tfrac{1}{\sqrt{c}}v), \quad u(0) \leq M\sqrt{V} \\ v' &\leq \gamma f(u + \tfrac{1}{\sqrt{c}}v), \quad v(0) \leq M \end{aligned} \tag{5.11}$$

where $\gamma \geq 1$, $V = \text{vol}(\Omega)$, $c = \frac{\gamma - 1}{V}$, $M > 0$, *and* $f(\theta)$ *is a nondecreasing function. Define* $N = \max\{M\sqrt{V}, M\} + \varepsilon$ *for some* $\varepsilon > 0$. *There is a* $\sigma > 0$ *such that* $u(t) \leq N$ *and* $v(t) \leq N$ *for all* $t \in [0, \sigma)$.

Proof. Let $\overline{u}(t)$ and $\overline{v}(t)$ be the solutions to

$$\begin{aligned} \overline{u}' &= \sqrt{\gamma V}\, f(\overline{u} + \tfrac{1}{\sqrt{c}}\overline{v}), \quad \overline{u}(0) = M\sqrt{V} \\ \overline{v}' &= \gamma f(\overline{u} + \tfrac{1}{\sqrt{c}}\overline{v}), \quad \overline{v}(0) = M \end{aligned}$$

for $t \geq 0$. The right-hand side components of (5.11) are quasimonotone nondecreasing since f is nondecreasing. By standard comparison results for systems of ordinary differential equations, $u(t) \leq \overline{u}(t)$ and $v(t) \leq \overline{v}(t)$. Since $\max\{\overline{u}(0), \overline{v}(0(\} < N$, there is a $\sigma > 0$ such that $\overline{u}(t) \leq N$ and $\overline{v}(t) \leq N$ for $t \in [0, \sigma)$. Consequently, $u(t) \leq N$ and $v(t) \leq N$ for $t \in [0, \sigma)$. □

Theorem 5.11 *If* $\theta_0 \in L^2(\Omega)$ *and* $\sup\{\theta_0(x) : x \in \Omega\} < \infty$, *then (5.10) has a unique solution* $\theta(t) \in L^2(\Omega)$, $t \in [0, \sigma)$, *for some* $\sigma > 0$.

Proof. Choose $M > 0$ such that $\|\theta_0\|_{L^2(\Omega)} < M$ and $\sup\{\theta_0(x) : x \in \Omega\} \leq M$. Set $N = \max\{M\sqrt{V}, M\} + \varepsilon$ for some $\varepsilon > 0$ where $V = \text{vol}(\Omega)$. For each $t \geq 0$, define $\theta_N(x,t) = \min\{\theta(x,t), N\}$ and $\eta_N(t) = \min\{\eta(t), N\}$. Consider the following auxiliary IBVP on the set E:

$$\begin{aligned} &\tfrac{d}{dt}(\theta, \eta) = \hat{A}(\theta, \eta) + \hat{B}_N(\theta, \eta), \quad (x,t) \in \Omega \times (0, \infty) \\ &(\theta(x,0), \eta(0)) = (\theta_0(x), 0), \quad x \in \overline{\Omega} \\ &\theta(x,t) = \eta(t), \quad (x,t) \in \partial\Omega \times (0, \infty) \end{aligned} \tag{5.12}$$

where $\hat{A}$ is defined as in Section 5.2 and where

$$\hat{B}_N(\theta, \eta) = \left(f(\theta_N - \tilde{\eta}_N), \frac{\gamma - 1}{V} \int_\Omega f(\theta_N - \tilde{\eta}_N)\, dy \right).$$

The operator $\hat{B}_N$ is globally Lipschitz continuous on E. Hence, (5.12) has a unique strong solution $\left(\hat{\theta}(x,t), \hat{\eta}(t)\right)$ defined for $t \in [0, \infty)$ (by the result [MAR, Thm.5.1, pg.355]).

We now prove that $\sup\{\theta(x,t) : x \in \Omega\}$ and $\eta(t)$ are bounded by N for $t \in [0, \sigma)$. This in turn implies that $(\hat{\theta}, \hat{\eta})$ is a solution of

$$\begin{aligned}
&\tfrac{d}{dt}(\theta, \eta) = \hat{A}(\theta, \eta) + \hat{B}(\theta, \eta), \quad (x,t) \in \Omega \times (0, \infty) \\
&(\theta(x, 0), \eta(0)) = (\theta_0(x), 0), \quad x \in \overline{\Omega} \\
&\theta(x,t) = \eta(t), \quad (x,t) \in \partial\Omega \times (0, \infty)
\end{aligned} \tag{5.13}$$

where $\hat{B}(\theta, \eta) = (f(\theta - \tilde{\eta}), \frac{\gamma-1}{V} \int_\Omega f(\theta - \tilde{\eta})\, dy)$. By Theorem 5.5 and the dissipativity of $\hat{A}$, the following inequalities are true:

$$\begin{aligned}
D^+\|(\theta, \eta)\| &\le \|\hat{B}_N(\theta, \eta)\| \\
&\le \sqrt{\gamma V}\, f(\sup\{\theta_N(x,t) : x \in \Omega\} + |\eta_N|), \\
D^+ \sup\{\theta_N(x,t) : x \in \Omega\} &\le \|\hat{B}_N(\theta, \eta)\|_\infty \\
&\le \gamma f(\sup\{\theta_N(x,t) : x \in \Omega\} + |\eta_N|)
\end{aligned} \tag{5.14}$$

where $\|(g, \xi)\|_\infty = \|g\|_\infty + |\xi|$. Moreover, it is easily seen that

$$|\eta_N| \le |\eta| \le \frac{1}{c} \|(\theta, \eta)\|. \tag{5.15}$$

Set $u(t) = \|(\theta, \eta)\|$ and $v(t) = \sup\{\theta(x,t) : x \in \Omega\}$. Using $\theta_N(x,t) \le \theta(x,t)$, and combining (5.14) and (5.15), we see that u and v are functions which satisfy (5.11). By Lemma 5.10, $u(t) \le N$ and $v(t) \le N$ for $t \in [0, \sigma)$, so $|\eta|$, $\|(\theta, \eta)\|$, and $\sup\{\theta(x,t) : x \in \Omega\}$ are all bounded by N for $t \in [0, \sigma)$. This proves that $(\hat{\theta}, \hat{\eta})$ is a strong solution of (5.13) on $[0, \sigma)$. An easy computation shows that $\theta(x,t) = \pi(\hat{\theta}, \hat{\eta}) = \theta(x,t) - \hat{\eta}(t)$ is a strong solution of (5.10). □

Corollary 5.12 *If $\theta_0 \in L^2(\Omega)$ and $\sup\{\theta_0(x) : x \in \Omega\} < \infty$, then IBVP (5.1)-(5.2) has a unique solution $\theta(x,t)$ on $\Omega \times [0, \sigma)$ for some $\sigma > 0$.*

Proof. By Theorem 5.11, $\pi(\hat{\theta}, \hat{\eta})$ is a strong solution of (4.10). Thus, it is a classical solution of (5.3)-(5.2) and a classical solution of (5.4)-(5.2). □

Theorem 5.13 *If $\theta_0 \in L^2(\Omega)$ and $\sup\{\theta_0(x) : x \in \Omega\} < \infty$, then IBVP (5.1)-(5.2) has a unique nonextendable classical solution $\theta(x,t)$ defined on a maximal interval $[0,\sigma)$ where*

$$\sigma = \infty, \text{ or, } \sigma < \infty \text{ and } \lim_{t\to\sigma^-} \sup\{\theta(x,t) : x \in \Omega\} = \infty.$$

Proof. Assume that $\sigma < \infty$ and $\theta(x,t) \leq N$ for all $(x,t) \in \Omega \times [0,\sigma)$. Let θ_N be the classical solution associated with (5.12) which exists on $\Omega \times [0,\infty)$. Our assumptions imply that $\theta_N = \theta$ on $\Omega \times [0,\sigma)$ and $\lim_{t\to\sigma^-} \theta(\cdot,t) = \theta_N(\cdot,\sigma)$. By Theorem 5.11 we can extend θ to a solution of (5.1)-(5.2) on $\Omega \times [0,\sigma+\varepsilon)$ for some $\varepsilon > 0$. This contradicts the maximality of $[0,\sigma)$. □

5.4 The Radially Symmetric Case

When $\Omega = B_1 \subset \mathbb{R}^n$ we can obtain additional information concerning initial-boundary value problem (5.1)-(5.2).

Theorem 5.14 *If $\theta(x,t)$ is the solution of IBVP (5.1)-(5.2) on $B_1 \times [0,\tau_0)$ with $\theta_0 \in \Gamma$, then $\theta(\cdot,t) \in \Gamma$ for all $t \in [0,\tau_0)$.*

Proof. By Theorem 5.7, the set Γ of nonnegative, nonincreasing, radially symmetric functions in $L^2(B_1)$ is invariant under the semigroup T. If $\theta(\cdot,t) \in \Gamma$, then

$$B[\theta(\cdot,t)] = f(\theta(\cdot,t)) + \frac{\gamma-1}{V}\int_{B_1} f(\theta(y,t))\,dy \in \Gamma.$$

In particular, $\theta(\cdot,t) + hB(\theta(\cdot,t)) \in \Gamma$ for all $h > 0$ because Γ is a convex cone.

For any $\tau \in (0,\tau_0)$, $\theta(\cdot,t)$ is a classical solution on $[0,\tau]$, so $\sup\{\theta(x,t) : (x,t) \in B_1 \times [0,\tau]\} \leq N < \infty$. This implies that $\theta(\cdot,t)$ is a solution of (5.1) lying in

$$F_N = \left\{g \in L^2(B_1) : \sup\{g(x) : x \in B_1\} \leq N\right\}$$

and the restriction of B to F_N is Lipschitz continuous. The result follows as a consequence of [MAR, Thm.2.1, pg.335]. □

Corollary 5.15 *Let $\theta_0 \in \Gamma$ and let ϕ be the solution to*

$$\phi_t - \Delta\phi = f(\phi) + \frac{\gamma-1}{\text{vol}(B_1)}\int_{B_1} f(\phi)\,dy, \quad (x,t) \in B_1 \times (0,\infty) \tag{5.16}$$

with initial-boundary conditions

$$\begin{aligned} &\phi(x,0) = \theta_0(x), \quad x \in \overline{B}_1 \\ &\phi(x,t) = 0, \quad (x,t) \in \partial B_1 \times (0,\infty); \end{aligned} \tag{5.17}$$

then $\phi(x,t) \geq \theta(x,t)$ *on their common interval of existence* $[0, \tau_0)$ *where* $\theta(x,t)$ *is the solution of IBVP (5.1)-(5.2).*

Proof. By Theorem 5.14, $\theta(\cdot,t) \geq 0$. Thus,

$$\frac{\gamma - 1}{\mathrm{vol}(B_1)} \int_{B_1} \Delta\theta \, dy = \frac{\gamma - 1}{\mathrm{vol}(B_1)} \int_{\partial B_1} \frac{\partial \theta}{\partial \nu} \, d\sigma \leq 0$$

for all $t \in [0, \tau_0)$. Consequently,

$$\begin{aligned} &\phi_t - \Delta\phi - f(\phi) - \tfrac{\gamma-1}{\mathrm{vol}(B_1)} \textstyle\int_{B_1} f(\phi)\, dy \\ &\quad \geq \theta_t - \Delta\theta - f(\theta) - \tfrac{\gamma-1}{\mathrm{vol}(B_1)} \textstyle\int_{B_1} f(\theta)\, dy. \end{aligned}$$

Using $f'(u) \geq 0$ and a maximum principle argument (*cf.* Theorem 4.1 and Corollary 4.2), one can prove that $\phi(x,t) \geq \theta(x,t)$ on their common interval of existence $[0, \tau_0)$. □

Theorem 5.16 *Let* $\theta_0 = 0$*; then the solution* $\theta(x,t)$ *of IBVP (5.1)- (5.2) is nondecreasing in* t *on its maximal interval of existence* $[0, \sigma)$ *for each* $x \in B_1$.

Proof. Choose any $\sigma_0 \in (0, \sigma)$. Since $\theta(x,t)$ can be thought of as a strong solution of (5.10), $\theta(\cdot,t) \leq N$ for some $N > 0$ and for all $t \in [0, \sigma_0]$. Also, $\theta(\cdot,t) \in \Gamma$ on $[0, \sigma)$. Thus, $\theta(\cdot,t)$ is also a solution to the associated problem (5.12) on $[0, \tau_0]$ and can be expressed by the Picard scheme $\{u_n\}_{n=0}^{\infty}$ where

$$u_0(t) = 0, \; u_{n+1}(t) = \int_0^t T(t-s) B(u_n(s)) \, ds \text{ for } n \geq 0,$$

and

$$\lim_{n \to \infty} u_n(t) = \theta(t),$$

since A is contractive and B is Lipschitz continuous.

Let $\theta_1, \theta_2, \theta \in \Gamma$ with $\theta_1 \leq \theta_2$. Note that $B : \Gamma \to \Gamma$ satisfies the following conditions:

$$B(\theta_1) \leq B(\theta_2) \quad \text{(monotonicity)}$$

and

$$B(\theta + \theta_1) - B(\theta_1) \leq B(\theta + \theta_2) - B(\theta_2) \quad \text{(convexity)}$$

with both sides of the above inequality elements of Γ.

Define

$$\begin{aligned} \Delta_n u(t) &= u_n(t) - u_{n-1}(t) \\ &= \textstyle\int_0^t T(t-s) \left[B(u_{n-1}(s)) - B(u_{n-2}(s)) \right] ds \end{aligned}$$

for $n \geq 2$ and

$$\Delta_n B(t) = B(u_n(t)) - B(u_{n-1}(t))$$

for $n \geq 1$; then $\Delta_{n+1}u(t) = \int_0^t T(t-s)\Delta_n B(s)\,ds$ and we claim that for any $n \geq 1$ the functions $u_n(t)$, $\Delta_n(t)$, and $\Delta_n B(t)$ are increasing in t with values in Γ. We prove this claim by induction.

For $n = 1$, clearly $u_1 \in \Gamma$ and $\Delta_1 u = u_1 \in \Gamma$. Since f is nondecreasing and since $u_1 \geq u_0 = 0$, $\Delta_1 B = B(u_1) - B(u_0) \in \Gamma$. One can also verify that u_1, $\Delta_1 u$, and $\Delta_1 B$ are increasing in t.

Assume that the claim is true for $n = 1, \ldots, k-1$; then $\Delta_k u(t) = \int_0^t T(t-s)\Delta_{k-1}B(s)\,ds \in \Gamma$ because $\Delta_{k-1}B(s) \in \Gamma$, Γ is invariant under T, and Γ is a closed convex cone. If $t_1 \leq t_2$, then since $\Delta_{k-1}B$ is nondecreasing we have

$$\begin{aligned}
\Delta_k u(t_1) &= \int_0^{t_1} T(t_1 - s)\Delta_{k-1}B(s)\,ds \\
&\leq \int_0^{t_1} T(t_1 - s)\Delta_{k-1}B(s + t_2 - t_1)\,ds \\
&= \int_{t_2 - t_1}^{t_2} T(t_2 - s)\Delta_{k-1}B(s)\,ds \\
&\leq \int_0^{t_2} T(t_2 - s)\Delta_{k-1}B(s)\,ds \\
&= \Delta_k u(t_2).
\end{aligned}$$

Thus, $u_k(t) = u_{k-1}(t) + \Delta_k u(t) \in \Gamma$ and is increasing since the same is true for $\Delta_k u(t)$ and $u_{k-1}(t)$. Finally, we have

$$\begin{aligned}
\Delta_k B(t) &= B(u_k(t)) - B(u_{k-1}(t)) \\
&= B(u_{k-1}(t) + \Delta_k u(t)) - B(u_{k-1}(t)) \\
&\in \Gamma
\end{aligned}$$

by the convexity of B. Furthermore, if $t_1 < t_2$, then the monotonicity of u_{k-1} and $\Delta_k u$ yields

$$\begin{aligned}
\Delta_k B(t_1) &= B(u_{k-1}(t_1) + \Delta_k u(t_1)) - B(u_{k-1}(t_1)) \\
&\leq B(u_{k-1}(t_2) + \Delta_k u(t_1)) - B(u_{k-1}(t_2)) \\
&\leq B(u_{k-1}(t_2) + \Delta_k u(t_2)) - B(u_{k-1}(t_2)) \\
&= \Delta_k B(t_2)
\end{aligned}$$

where the convexity of B was used at the first inequality and the monotonicity of B was used at the second inequality.

This shows that $\theta(t)$ is a limit of an increasing sequence $\{u_n\}_{n=0}^{\infty}$ of t-increasing functions. Thus, $\theta(t)$ is itself an increasing function. □

Corollary 5.17 *If $\theta_0(x) \equiv 0$, then the solution $\theta(x,t)$ of IBVP (5.1)- (5.2) is an upper solution for the solution $\psi(x,t)$ of IBVP (1.28)-(1.29).*

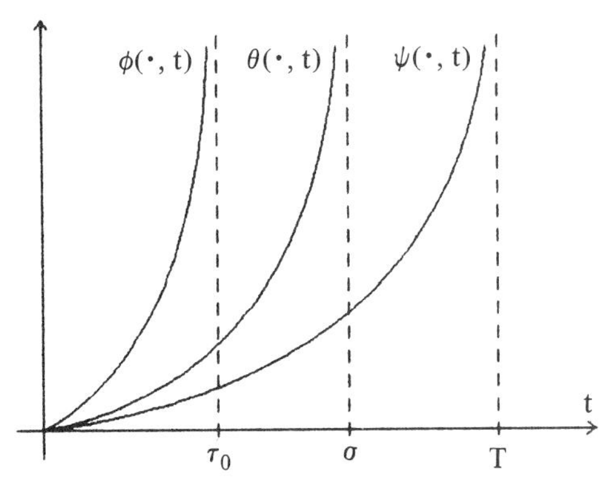

Figure 5.1.

Proof. By Theorem 5.16, the solution $\theta(x,t)$ of (5.1)-(5.2) satisfies $\theta_t(x,t) \geq 0$. Thus,

$$f(\theta) + \frac{\gamma - 1}{\text{vol}(B_1)} \int_{B_1} f(\theta(y,t))\, dy \geq f(\theta)$$

and the result is immediate from standard comparison results. □

Corollary 5.17 tells us that the temperature for an ideal gas is always greater than that for a solid fuel. Hence, a gas explodes sooner than a solid in the same container. Physically, this can be explained by the additional generation of heat due to the compression of the gas.

For $\Omega = B_1 \subset \mathbb{R}^n$ and $\delta > \delta_{FK}$, the solution ψ of IBVP (1.28)-(1.29) blows up in finite time T. Figure 5.1 illustrates the comparisons where ϕ is the solution of IBVP (5.16)- (5.17) and ψ is the solution of IBVP (1.28)-(1.29).

Table 5.1 gives a comparison of blowup times for the three problems: IBVP (5.16)-(5.17), IBVP (1.28)-(1.29), and IBVP (1.39)-(1.40) (for $a = 1$). The table uses $\Omega = (-1, 1)$ and $\gamma = 1.4$.

Table 5.1.

δ	τ_0	σ	T
0.91	1.755	6.123	7.940
1.00	1.401	2.732	3.537
2.00	0.454	0.528	0.680
2.47	0.347	0.390	0.502
20.00	0.037	0.038	0.050
50.00	0.0147	0.0148	0.020

Table 5.2.

		σ		
δ	T	$\gamma = 1.1$	$\gamma = 1.4$	$\gamma = 2.0$
3.7	0.876	0.854	0.802	0.736
4.0	0.601	0.580	0.532	0.473
6.7	0.187	0.179	0.160	0.138
20.0	0.0503	0.0478	0.0418	0.0336
50.0	0.0200	0.0188	0.0158	0.0120

As the gas constant γ varies, we can numerically compute σ. Table 5.2 uses $\Omega = B_1 \subset \mathbb{R}^3$.

Note that Theorem 5.16 and Corollary 5.17 are proved for $\theta_0 \equiv 0$. An open problem is the following: Is the result true for any $\theta_0 \in \Gamma$? The key idea is to prove that $\theta_1 \geq \theta_0$ so that the case $n = 1$ is true in the induction proof.

Theorem 5.18 *Assume that $\overline{\theta}_1, \overline{\theta}_2 \in \Gamma$ satisfy $\overline{\theta}_2 - \overline{\theta}_1 \in \Gamma$ and suppose that $0 < \delta_1 \leq \delta_2$. Let $\theta_i(x,t)$ be a solution to IBVP (5.1)-(5.2) with θ_0 and δ replaced by $\overline{\theta}_i$ and δ_i for $i = 1, 2$; then $\theta_1(x,t) \leq \theta_2(x,t)$ for all (x,t) in their common domain of existence.*

Proof. Define the set $\Phi = \{(\theta_1, \theta_2) \in \Gamma \times \Gamma : \theta_2 - \theta_1 \in \Gamma\}$. This set is a closed convex cone in $F \times F$. We claim that Φ is invariant under the flow generated by $(A + B_1) \times (A + B_2)$ where B_i $(i = 1, 2)$ are defined as in (5.10) replacing δ by δ_i. Theorem 5.7 implies that Φ is invariant under the flow induced by $A \times A$. If $(\theta_1, \theta_2) \in \Phi$, then $(B_1\theta_1, B_2\theta_2) \in \Phi$. Thus, Φ is invariant and by Corollary 5.8, $\theta_2(x,t) \geq \theta_1(x,t)$. $\square$

If $\theta_0 \in \Gamma$, then the solutions of initial-boundary value problem (5.1)-(5.2) depend monotonically on the parameter δ. As a consequence, we have for the subcritical case $\delta < \delta_{FK}$:

Corollary 5.19 *Assume that BVP (1.30)-(1.31) has a solution $\psi(x)$ on the unit ball B_1; then the solution $\theta(x,t)$ of IBVP (1.28)-(1.29) exists on $B_1 \times [0, \infty)$ with $\theta(x,t) \leq \psi(x)$.*

5.5 Blowup: Where?

We consider the partial differential equation

$$u_t - \Delta u = f(u) + g(t), \quad (x,t) \in \Omega \times (0,T) \tag{5.18}$$

with initial-boundary conditions

$$\begin{aligned} u(x,0) &= \phi(x), \quad x \in \overline{\Omega} \\ u(x,t) &= 0, \quad (x,t) \in \partial\Omega \times (0,T) \end{aligned} \tag{5.19}$$

where $\Omega = \{x \in \mathbb{R}^n : |x| < R\}$. We assume that $\phi \in C^2(\overline{\Omega}, [0,\infty))$, $\Delta\phi + f(\phi) \geq 0$ for $x \in \Omega$, $\phi(x) = 0$ for $x \in \partial\Omega$, ϕ is radially symmetric and radially decreasing.

We also assume that $f \in C^2(\mathbb{R}, [0,\infty))$, $f(u) > 0$ for $u > 0$, $f'(u) \geq 0$, $f''(u) \geq 0$, and $\int^\infty \frac{du}{f(u)} < \infty$. We choose $g(t)$ such that either $g \in C^1$, $g \geq 0$, and $g' \geq 0$, or, $g(t) = \frac{K}{V}\int_\Omega u_t\,dy$ with $K \in (0,1)$ and $\phi(x) \equiv 0$.

By the results of Section 5.4 there is a unique solution $u(x,t)$ for $(x,t) \in \overline{\Omega} \times [0,\sigma)$ such that $u(x,t) \geq 0$ and $u_t(x,t) \geq 0$. Moreover, $u(\cdot,t)$ is radially symmetric and radially decreasing. The function $U(t) = \sup\{u(x,t) : x \in \Omega\} = u(0,t)$ is a nondecreasing function.

Let $[0,T]$ be the maximal interval of existence for the solution $u(x,t)$ to (5.18)-(5.19). If $T < \infty$, then $U(T^-) = \infty$ by Theorem 5.13. We will assume that $T < \infty$ so that $u(x,t)$ blows up in finite time. As in Section 3.3, we define a point $x \in \Omega$ to be a *blowup point* for (5.18)-(5.19) if there is a sequence $\{(x_m, t_m)\}_{m=0}^\infty$ such that $t_m \to T^-$, $x_m \to x$, and $u(x_m, t_m) \to \infty$ as $m \to \infty$.

Theorem 5.20 *If $\int_0^T g(t)\,dt = \infty$, then the solution $u(x,t)$ of (5.18)-(5.19) satisfies $\lim_{t\to T^-} u(x,t) = \infty$ for all $x \in \Omega$. Thus, blowup occurs everywhere in Ω.*

Proof. Choose $\overline{x} \in \Omega$ and define $\rho = R - |\overline{x}|$. On the ball $B_\rho(\overline{x}) \subset \Omega$, the solution $u(x,t)$ is an upper solution for

$$\begin{aligned} &v_t = \Delta v + g(t), \quad (x,t) \in B_\rho(\overline{x}) \times (0,T) \\ &v(x,0) = 0, \quad x \in B_\rho(\overline{x}) \\ &v(x,t) = 0, \quad x \in \partial B_\rho(\overline{x}) \times (0,T). \end{aligned}$$

The solution $v(x,t)$ at $(\overline{x},t)$ can be expressed as

$$\begin{aligned} v(\overline{x},t) &= \int_0^t \int_{B_\rho(\overline{x})} G(\overline{x},y,t-s) g(s)\,dy\,ds \\ &\geq \int_0^t g(s) \int_{B_\rho(\overline{x})} G(\overline{x},y,T-0)\,dy\,ds \\ &\geq K(\rho) \int_0^t g(s)\,ds \end{aligned}$$

where $K(\rho) = \int_{B_\rho(\overline{x})} G(\overline{x},y,T)\,dy$. As $t \to T^-$, $v(\overline{x},t) \to \infty$ by our hypothesis on the integral of $g(s)$. Since $v(x,t) \leq u(x,t)$ on $B_\rho(\overline{x})$, we have $u(\overline{x},t) \to \infty$. Since $\overline{x}$ was arbitrary, the solution $u(x,t)$ blows up for all $x \in \Omega$. □

Theorem 5.21 *Let $g(t) = \frac{K}{\text{vol}(\Omega)} \int_\Omega u_t(x,t)\,dx$ for $0 \le K < 1$. If (5.18)-(5.19) has a blowup point $\overline{x} \neq 0$, then the solution $u(x,t)$ blows up everywhere in Ω.*

Proof. Observe that

$$\int_0^t g(s)\,ds = \frac{K}{\text{vol}(\Omega)} \int_0^t \int_\Omega u_t(x,s)\,dx\,ds = \frac{K}{\text{vol}(\Omega)} \int_\Omega u(x,t)\,dx$$

since $u(x,0) = \phi(x) \equiv 0$. By radial monotonicity of u,

$$\int_\Omega u(x,t)\,dx \ge \int_{|x|\le|\overline{x}|} u(x,t)\,dx \ge \text{vol}\left(B_{|\overline{x}|}(0)\right) u(\overline{x},t)$$

so that if $\lim_{t\to T^-} u(\overline{x},t) = \infty$ for some $\overline{x} \neq 0$, then $\int_0^T g(s)\,ds = \infty$. By Theorem 5.20 blowup must occur everywhere in Ω. □

Theorem 5.22 *If $f(u) = e^u$ and $g(t) = \frac{K}{\text{vol}(\Omega)} \int_\Omega u_t\,dx$ for $K \in (0,1)$, then the solution $u(x,t)$ of (5.18)-(5.19) blows up only at $x = 0$.*

Proof. The proof is similar to that of Theorem 3.16. Define the function

$$J(r,t) = r^{n-1}u_r(r,t) + \varepsilon r^n F(u(r,t),t)$$

where $F(u,t) = \exp(\alpha(u - G(t)))$ and $G(t) = \int_0^t g(s)\,ds$ for $\alpha \in (0,1)$. It can be shown that J satisfies

$$J_t + \frac{n-1}{r}J_r - J_{rr} - (f' - 2\varepsilon F_u)J \le -\varepsilon r^n(f'F - F_u f - 2\varepsilon F F_u) + \varepsilon r^n(F_u g + F_t).$$

If

$$f'F - F_u f \ge 2\varepsilon F F_u \quad \text{and} \quad F_u g + F_t \le 0, \tag{5.20}$$

then

$$J_t + \frac{n-1}{r}J_r - J_{rr} - (f' - 2\varepsilon F_u)J \le 0$$

for $(r,0) \in (0,R) \times (0,T)$. The second inequality in (5.20) is immediate from our definition of $F(u,t)$. The first inequality in (5.20) is valid for $\varepsilon \le (1-\alpha)/2\alpha$.

Note that $J(0,t) = 0$. Next observe that

$$\begin{aligned} J_r(R,t) &= -R^{n-1}\left[f(0) + g(t) - \varepsilon n e^{-\alpha G(t)}\right] \\ &\le R^{n-1}[\varepsilon n - 1 - g(t)] \\ &< 0 \end{aligned}$$

if $\varepsilon < 1/n$. Finally, $J(r,\eta) < 0$ for η sufficiently close to 0, just as in the proof of Theorem 3.16. By the maximum principle, $J(r,t) \leq 0$ for $(r,t) \in [0,R] \times [0,T)$ and so

$$r^{n-1} u_r \leq -\varepsilon r^n e^{\alpha(u-G(t))}.$$

By integration we obtain

$$u(r,t) \leq \frac{2}{\alpha}\ln(\frac{1}{r}) - \frac{1}{\alpha}\ln(\frac{\alpha\varepsilon}{2}) + G(t). \tag{5.21}$$

Integrating over Ω and recalling from Theorem 5.21 that

$$G(t) \leq \frac{K}{\text{vol}(\Omega)} \int_\Omega u(x,t)\,dx$$

we obtain

$$\int_\Omega u(x,t)\,dx \leq \int_\Omega \left[\frac{2}{\alpha}\ln\left(\frac{1}{|x|}\right) - \frac{1}{\alpha}\ln\left(\frac{\alpha\varepsilon}{2}\right)\right] dx + K\int_\Omega u(x,t)\,dx$$

or

$$(1-K)\int_\Omega u(x,t)\,dx \leq \int_\Omega \left[\frac{2}{\alpha}\ln\left(\frac{1}{|x|}\right) - \frac{1}{\alpha}\ln\left(\frac{\alpha\varepsilon}{2}\right)\right] dx < \infty$$

and by Theorem 5.21 blowup can occur only at a single point provided $K \in (0,1)$. □

Corollary 5.23 *If $f(u) = e^u$ and $\int_0^T g(s)\,ds < \infty$, then blowup occurs only at $x = 0$.*

Proof. Note that in the proof of Theorem 5.22, the construction leading up to (5.21) is valid for an arbitrary $g(t)$. As long as $G(T)$ is finite, (5.21) implies that $u(r,t)$ is finite for $r \neq 0$. □

The following theorem is a technical result which allows us to obtain lower bounds on the solution $u(x,t)$. The proof is based on ideas from [FRI2, Thm. 3.1].

Theorem 5.24 *Assume that $\int_0^\infty f(u)\,du = \infty$. Let $u(x,t)$ be a solution to (5.18)-(5.19) which blows up only at $x = 0$; then there exists a $t^* \in (0,T)$ such that*

$$|\nabla u(x,t)|^2 \leq 2\left[-F(u(x,t)) + F(U(t)) + Lf(U(t))\right] \tag{5.22}$$

for $t \in (t^*,T)$ where

$$F(w) = \int_0^w f(u)\,du, \quad L = \int_0^T g(s)\,ds < \infty,$$

and

$$U(t) = u(0,t) = \max_{x\in\overline{\Omega}} u(x,t).$$

Proof. Note that $L < \infty$ by the assumptions of single-point blowup and Theorem 5.20. Since u blows up only at the origin, both u and ∇u are uniformly bounded on the parabolic boundary of the cylinder

$$Q = \{(x,t) : |x| \leq R/2,\, 0 \leq t < T\}.$$

Consequently,

$$\max_{(x,t)\in\partial Q} \left[\frac{1}{2}|\nabla u(x,t)|^2 + F(u(x,t))\right] =: M < \infty.$$

Since $U(t) \uparrow \infty$ as $t \to \infty$ and since $F(w) \to \infty$ as $w \to \infty$, there is a $t^* < T$ such that $F(U(t)) > M$ for all $t \in [t^*, T)$.

For any $\bar{t} \in [t^*, T)$ define the function

$$J(x,t) = \frac{1}{2}|\nabla u(x,t)|^2 + F(u(x,t)) - F(U(\bar{t})) - f(U(\bar{t})) \int_0^t g(s)\, ds.$$

We will show by a maximum principle that $J(x,t) \leq 0$ on the cylinder $\{(x,t) : |x| < R/2,\, 0 \leq t \leq \bar{t}\}$. This condition on J implies the bound (5.22).

On the set ∂Q, we have

$$J(x,t) \leq M - F(U(\bar{t})) - f(U(\bar{t})) \int_0^t g(s)\, ds < 0.$$

Moreover, for $x = 0$ and $t \in [0, \bar{t})$ we get

$$J(0,t) \leq F(u(x,t)) - F(U(\bar{t})) \leq 0.$$

It can be shown that

$$J_t =$$

$$\nabla u \bullet \nabla(\Delta u) + f'(u)|\nabla u|^2 + f(u)\Delta u + f^2(u) + f(u)g(t) - f(U(\bar{t}))g(t),$$

$$\nabla J = (\Delta u + f(u))\nabla u, \quad \text{and}$$

$$\Delta J = (\Delta u)^2 + \nabla u \bullet \nabla(\Delta u) + f'(u)|\nabla u|^2 + f(u)\Delta u.$$

Combining this with the identity

$$|\nabla J - (\Delta u)\nabla u|^2 = |\nabla u|^2(\Delta u)^2 + \nabla J \bullet [\nabla J - 2(\Delta u)\nabla u] = f^2(u)|\nabla u|^2,$$

we obtain

$$J_t - \Delta J - \frac{\nabla J \bullet [\nabla J - 2(\Delta u)\nabla u]}{|\nabla u|^2} = [f(u(x,t)) - f(U(\bar{t}))]g(t) \leq 0.$$

Since $\nabla u = 0$ only at $x = 0$, the maximum principle implies that $J(x,t) \leq 0$ on $\{(x,t) : |x| < R/2,\, 0 \leq t \leq \bar{t}\}$. In particular, $J(x,\bar{t}) \leq 0$ and the theorem is proved. □

Theorem 5.25 *Let $f(u) = (u+\lambda)^p$ for $\lambda \geq 0$ and $1 < p < 1 + 2/n$. Let $g(t) = \frac{K}{\text{vol}(\Omega)} \int_\Omega u_t(x,t)dx$ with $K \in (0,1)$; then the solution $u(x,t)$ to IBVP (5.18)-(5.19) blows up everywhere in Ω.*

Proof. If the conclusion were false, then single-point blowup occurs only at $x = 0$. From the fact that u is radially symmetric, we have by Theorem 5.24

$$|u_r(r,t)|^2 \leq 2f(U(t))[U(t) - u(r,t) + L].$$

Thus, we have

$$\int_0^r \frac{-u_r(r,t)}{[U(t) - u(r,t) + L]^{1/2}}\, dr \leq \int_0^r \left[2f(U(t))\right]^{1/2} dr$$

which implies

$$u(r,t) \geq U(t) - L - f(U(t))r^2.$$

For

$$r_1 = \left[\frac{\frac{U(t)}{2} - L}{f(U(t))}\right]^{1/2}$$

we have $u(r,t) \geq \frac{U(t)}{2}$ for $r \leq r_1$. Define ω_n to be the surface area of the unit n-dimensional ball; then

$$\begin{aligned}
\int_\Omega u(x,t)\, dx &= \omega_n \int_0^R r^{n-1} u(r,t)\, dr \\
&\geq \omega_n \int_0^{r_1} \tfrac{1}{2} r^{n-1} U(t)\, dr \\
&= \frac{\omega_n U(t)}{2n} \left[\frac{\frac{U(t)}{2} - L}{f(U(t))}\right]^{n/2}
\end{aligned}$$

Since $f(s) = \mathbf{o}(s^{1+2/n})$ as $s \to \infty$, $\int_\Omega u(x,t)\, dx = \infty$ as $t \to T^-$ and hence

$$\lim_{t \to T^-} \int_0^t g(s)\, ds = \lim_{t \to T^-} \left[\frac{K}{\text{vol}(\Omega)} \int_\Omega u(x,t)\, dx\right] = \infty$$

which is a contradiction to Theorem 5.20. Thus, the solution $u(x,t)$ must blow up everywhere in the set Ω. □

Theorem 5.26 *If $f(u) = (u+\lambda)^p$ with $p > 1 + 2/n$ and*

$$g(t) = \frac{K}{\text{vol}(\Omega)} \int_\Omega u_t(x,t)\, dx,$$

then the solution of (5.18)-(5.19) blows up only at $x = 0$ as long as $K \geq 0$ is sufficiently small.

Proof. The idea of the proof is exactly the same as that of Theorem 5.22. Define the function

$$J(r,t) = r^{n-1}u_r(r,t) + \varepsilon r^n F(u(r,t),t)$$

where $F(u,t) = (u+\mu)^q \exp(-\alpha G(t))$, $\varepsilon > 0$, $1 + \frac{2}{n} < q < p$, $\mu \geq \lambda$, $\mu > 0$, and $\alpha > 0$. As in Theorem 5.22, we can show via a maximum principle that $J(r,t) \leq 0$ on $(0,R) \times (0,T)$ so that

$$r^{n-1}u_r(r,t) \leq -\varepsilon r^n (u(r,t)+\mu)^q e^{-\alpha G(t)}.$$

An integration gives us

$$u(r,t) \leq \left[\frac{2e^{\alpha G(t)}}{\varepsilon(q-1)r^2}\right]^{\frac{1}{q-1}} \tag{5.23}$$

Integrating (5.23) over Ω gives us an inequality of the form

$$G(t) \leq KAe^{BG(t)}$$

where $A > 0$, $B > 0$, and $K \in (0,1)$. Choose K sufficiently small so that $KA < \frac{1}{Be}$; then $G(0) = 0$, $G(t)\exp(-BG(t)) \leq KA < \frac{1}{Be}$, and the continuity of $G(t)$ imply $G(t)$ is bounded for all $t \geq 0$. Consequently, the only blowup point is at $x = 0$. □

Corollary 5.27 *If $f(u) = (u+\lambda)^p$ for $p > 1$ and if $\int_0^T g(s)\,ds < \infty$, then the solution $u(x,t)$ of (5.18)-(5.19) blows up only at $x = 0$.*

Proof. In Theorem 5.26 we had derived equation (5.22) independently of the choice of $g(s)$. From this inequality, $u(x,t)$ is bounded as long as $G(t)$ is bounded and $x \neq 0$. □

5.6 A Nondiffusive Reactive Model

For an arbitrary container $\Omega \subset \mathbb{R}^n$, the nondiffusive reactive Euler model (1.41)-(1.40a) can be written as

$$\phi_t = \delta e^{\phi} + \frac{\gamma-1}{\gamma}\frac{1}{\mathrm{vol}(\Omega)}\int_\Omega \phi_t(x,t)\,dx, \quad (x,t) \in \Omega \times (0,\infty) \tag{5.24}$$

with initial data

$$\phi(x,0) = \phi_0(x), \quad x \in \Omega \tag{5.25}$$

assuming $\phi_0(x)$ is continuous and bounded on Ω. By integrating over Ω, we see that (5.24) is equivalent to

$$\phi_t = \delta e^{\phi} + \beta \int_\Omega e^{\phi(x)}\,dx \tag{5.26}$$

where $\beta = \frac{(\gamma-1)\delta}{\text{vol}(\Omega)}$. The IBVP (5.26)- (5.24) has a unique nonextendable solution $\phi(x,t)$ on $\overline{\Omega}\times[0,\sigma)$ with $\sigma = \infty$, or, $\sigma < \infty$ and $\lim_{t\to\sigma^-} \sup\{\phi(x,t) : x \in \Omega\} = \infty$.

The initial value problem

$$a_t = \delta e^a, \quad (x,t) \in \Omega \times (0,T) \tag{5.27}$$

$$a(x,0) = \phi_0(x), \quad x \in \overline{\Omega} \tag{5.28}$$

has the explicit solution

$$a(x,t) = -\ln\left[e^{-\phi_0(x)} - \delta t\right] \tag{5.29}$$

which blows up in finite time $T = \frac{1}{\delta}\exp(-\phi_0(x_m))$ where x_m is any point in Ω at which $\phi_0(x)$ attains its absolute maximum. Since $a(x,t)$ is a lower solution for (5.25)-(5.24), the solution $\phi(x,t)$ of (5.25)-(5.24) satisfies

$$\phi(x,t) \geq -\ln\left[e^{-\phi_0(x)} - \delta t\right]$$

and hence $\phi(x,t)$ blows up in finite time σ with $\sigma \leq T$.

To get more information about $\phi(x,t)$, we consider the implicit representation

$$\phi(x,t) = a(x,\tau(t)) + B(\tau(t)) \tag{5.30}$$

where $a(x,\tau)$ is a solution of (5.27)-(5.28) and $\tau(t)$, $B(\tau)$ are scalar functions to be determined. As given in (5.30), $\phi(x,t)$ is a solution of (5.26) if and only if

$$\tau' = e^{B(\tau)}, \quad \tau(0) = 0 \tag{5.31}$$

and

$$B' = \beta \int_\Omega e^{a(x,\tau)}\,dx = \beta \int_\Omega \left[e^{-\phi_0(x)} - \delta\tau\right]^{-1} dx, \quad B(0) = 0. \tag{5.32}$$

The system (5.31)-(5.32) is weakly coupled, so by integrating (5.32) from 0 to τ, we have

$$B(\tau) = \frac{\beta}{\delta}\int_\Omega [a(x,\tau) - \phi_0(x)]\,dx = \frac{\beta}{\delta}\int_\Omega \ln\left(\frac{e^{-\phi_0(x)}}{e^{-\phi_0(x)} - \delta\tau}\right) dx. \tag{5.33}$$

Thus, τ satisfies

$$\tau' = \exp\left[\frac{\beta}{\delta}\int_\Omega \ln\left(\frac{e^{-\phi_0(x)}}{e^{-\phi_0(x)} - \delta\tau}\right) dx\right], \quad \tau(0) = 0 \tag{5.34}$$

which can be solved by quadrature. Thus, $\phi(x,t) = a(x,\tau(t)) + B(\tau(t))$ is the solution of (5.26)-(5.25) where $\tau(t)$ solves (5.34), $a(x,\tau)$ is the solution of (5.27)-(5.28), and $B(\tau)$ is given by (5.32).

From (5.30) we have

Theorem 5.28 *The value σ is the blowup time for the solution $\phi(x,t)$ of (5.24)- (5.25) if and only if $\tau(\sigma) = T$ is the blowup time for the solution $a(x,\tau)$ of (5.27)-(5.28).*

If $\gamma > 1$, then $\tau'(t) > 1$ from (5.34) for $t > 0$, and hence, $\tau(t)$ is strictly increasing with $\tau(t) > t$ for $t > 0$. Thus,

Corollary 5.29 *The blowup time σ is given by*

$$\sigma = \tau^{-1}\left(\frac{1}{\delta}e^{-\phi_0(x_m)}\right)$$

where x_m is any point of Ω at which ϕ_0 has an absolute maximum.

From (5.30) and (5.33), observe that $\phi(x,t)$ blows up at those points x_m at which $\phi_0(x)$ has its absolute maximum, provided $B(\tau(\sigma)) < \infty$. This is true if and only if $\int_\Omega a(x,\tau(\sigma))\,dx < \infty$ which in turn is true provided $\int_\Omega \ln[\exp(-\phi_0(x)) - \exp(-\phi_0(x_m))]\,dx > -\infty$. Thus,

Theorem 5.30 *The solution $\phi(x,t)$ of (5.24)-(5.25) blows up only at those points x_m of Ω at which $\phi_0(x)$ has its absolute maximum if and only if*

$$\int_\Omega \ln\left[e^{-\phi_0(x)} - e^{-\phi_0(x_m)}\right]\,dx > -\infty. \tag{5.35}$$

Similarly, we observe that $\phi(x,t)$ blows up everywhere in Ω at σ if and only if $B(\tau(\sigma)) = \infty$. Thus,

Theorem 5.31 *The solution $\phi(x,t)$ blows up everywhere in Ω at σ if and only if*

$$\int_\Omega \ln\left[e^{-\phi_0(x)} - e^{-\phi_0(x_m)}\right]\,dx = -\infty. \tag{5.36}$$

The integral of (5.35) is finite if there is at most a finite number of critical points $x_m \in \Omega$ at which ϕ_0 has an absolute maximum and if at each x_m, $\phi_0(x)$ is strictly concave down and analytic in a neighborhood of x_m. In this case, blowup occurs only at those x_m at which ϕ_0 has an absolute maximum.

On the other hand,, if ϕ_0 is too flat in a neighborhood of a point x_m, then blowup occurs everywhere.

5.7 Comments

The first gaseous ignition model was developed by Kassoy and Poland [KAS5]. This model (1.39)-(1.40) was initially analyzed by Bebernes and Bressan [BEB5]. Several problems remain open. For example, can one compare the gaseous ignition model with the solid fuel model for nonzero initial data?

The semigroup theory used to prove existence and the invariance of certain sets is standard and can be found in such monographs as Martin [MAR] and Yosida [YOS].

The discussion of where blowup occurs is based on the paper [BEB12] which in turn draws from the seminal ideas of Friedman and McLeod [FRI2]. There are many open problems for arbitrary domains Ω and nonzero initial data.

Theorem 5.26 is proved only for $K > 0$ sufficiently small and $f(u) = (u+\lambda)^p$, $p > 1+2/n$. We conjecture that it is also true for any $K \in (0,1)$.

The nondiffusive model (5.24)-(5.25) was first considered in [BEB13]. A formal asymptotic description of how the blowup hot spot develops is also given there, but a rigorous analysis has not been carried out. This should not be difficult to do. It is interesting to note that for this model without diffusion that the blowup singularity is strongly dependent on the shape of the initial temperature profile.

6

Conservation Systems for Reactive Gases

In one space dimension, the conservation laws for reactive gases can be expressed as

$$u_t + F(u)_x = Bu_{xx} + G(u, u_x), \quad (x,t) \in \Omega \times (0,T) \subset \mathbb{R} \times \mathbb{R} \tag{6.1}$$

where the solutions u are vector-valued functions of (x,t), and where B is a positive semidefinite matrix which will be referred to as the *viscosity matrix.*

In Section 6.1 we will consider a special case of (6.1) where there is no reactive term ($G \equiv 0$) and where $F(u) = \nabla\Phi(u)$ for some function $\Phi \in C^2(\mathbb{R}^n, \mathbb{R})$ with $\Phi(0) = 0$. The boundary conditions are assumed to be Dirichlet. If $\langle v, Bv \rangle \geq \varepsilon > 0$ for some $\varepsilon > 0$ and for all $v \neq 0$, and if the Hessian matrix of $\Phi(u)$ essentially grows slower than $|u|^2$, we prove that there is a solution u which exists globally such that $u \to 0$ as $t \to \infty$ uniformly on Ω.

A nondiffusive-reactive Euler model is analyzed in Section 6.2. This model is a special case of (6.1) where $B \equiv 0$ and where $G = G(u)$ contains exponential nonlinearities. The function $F(u)$ is a linear function. It is shown that solutions to the (hyperbolic) initial value problem blow up in finite time. At a blowup point, the shape of solutions as the blowup time is reached is determined when $F \equiv 0$.

The remainder of the chapter is devoted to the analysis of the full one- dimensional gas model. The model is a special case of (6.1) where $u = (\rho, v, \theta, z)$ is the state of the system. The components of u represent density ρ, velocity v, temperature θ, and concentration z. The nonlinearities $G = G(u, u_x)$ and $F = F(u)$ are complicated functions. The viscosity matrix $B = \text{diag}\{0, \frac{\lambda_1}{\rho}, \frac{\lambda_2}{\rho}, \lambda_3\}$ and is positive semidefinite. For boundary conditions representing a thermally insulated container Ω, and under appropriate smoothness conditions, there is a unique classical solution u which exists for all time $t \geq 0$.

6.1 A Nonreactive Model

Let Ω be any bounded open interval of $\mathbb{R}$. Without loss of generality in the following development, we will choose $\Omega = (0,1)$. Let $\Pi_T = \Omega \times (0,T)$. We will consider here the particular gradient system

$$u_t + [\nabla\Phi(u)]_x = Bu_{xx}, \quad (x,t) \in \Pi_T \tag{6.2}$$

where $\Phi : \mathbb{R}^n \to \mathbb{R}$ is C^2-smooth and where $\Phi(0) = 0$. The initial-boundary conditions for (6.2) will be

$$\begin{aligned} &u(x,0) = u_0(x), \quad x \in \overline{\Omega} \\ &u(x,t) = 0, \quad (x,t) \in \partial\Omega \times [0,T) \end{aligned} \tag{6.3}$$

where $u_0 \in L^2(\Omega)$. Let $\langle \cdot, \cdot \rangle$ be the usual inner product on $\mathbb{R}^n$. We assume that $\langle v, Bv \rangle \geq \varepsilon > 0$ for some $\varepsilon > 0$ and for all $v \neq 0$.

Local existence for (6.2)-(6.3) is immediate. For global existence, one needs some kind of growth estimate on the nonlinearity $\Phi(u)$. Let $\Phi''(u)$ be the Hessian matrix of $\Phi(u)$ with $\|\cdot\|$ any convenient norm. Define $\beta(M) = \max\{\|\Phi''(u)\| : u \leq m\}$. Assume

$$\lim_{M\to\infty} \frac{M^2}{1+\beta(M)} = \infty. \tag{6.4}$$

This condition implies that $\|\Phi''(u)\|$ grows slower than $|u|^2$ and that $|\Phi(u)|$ grows slower than $|u|^4$; see [KAN].

Lemma 6.1 *Let $u(x,t)$ be a solution to initial-boundary value problem (6.2)-(6.3); then*

$$\|u(\cdot,T)\|^2_{L^2(\Omega)} \leq 2C_0 \quad \textit{and} \quad \|u_x\|^2_{L^2(\Pi_T)} \leq \frac{C_0}{\varepsilon}$$

for all $T \geq 0$ where $C_0 = \frac{1}{2}\|u_0\|^2_{L^2(\Omega)}$.

Proof. We use energy estimates for the norm of u. Taking the inner product of u with (6.2), we have

$$\langle u, Bu_{xx}\rangle = \langle u, u_t\rangle + \langle u, \nabla\Phi(u)_x\rangle = \frac{1}{2}\frac{\partial |u|^2}{\partial t} + \frac{\partial}{\partial x}[\langle u, \nabla\Phi(u)\rangle - \Phi(u)].$$

Integrating this equation over Π_T, we obtain

$$\begin{aligned} &\textstyle\iint_{\Pi_T} \langle u, Bu_{xx}\rangle\, dx\, dt \\ &= \textstyle\iint_{\Pi_T} \frac{1}{2}\frac{\partial |u|^2}{\partial t}\, dx\, dt + \iint_{\Pi_T} \frac{\partial}{\partial x}[\langle u, \nabla\Phi(u)\rangle - \Phi(u)]\, dx\, dt \\ &= \textstyle\frac{1}{2}\int_0^1 |u(x,T)|^2 dx - \frac{1}{2}\int_0^1 |u_0(x)|^2 dx + \int_0^T [\langle u, \nabla\Phi(u)\rangle - \Phi(u)]\Big|_0^1 dt \\ &= \textstyle\frac{1}{2}\int_0^1 |u(x,T)|^2 dx - \frac{1}{2}\int_0^1 |u_0(x)|^2 dx \end{aligned}$$

where we have used the boundary conditions in (6.3) and $\Phi(0) = 0$. Integrating the left-hand side by parts, we have

$$\iint_{\Pi_T} \langle u, Bu_{xx}\rangle\, dx\, dt = -\iint_{\Pi_T} \langle u_x, Bu_x\rangle\, dx\, dt \leq -\varepsilon \iint_{\Pi_T} |u_x|^2 dx\, dt.$$

Consequently, we have the inequality

$$\frac{1}{2}\int_0^1 |u(x,T)|^2 dx + \varepsilon \iint_{\Pi_T} |u_x|^2 dx\, dt \le \frac{1}{2}\int_0^1 |u_0(x)|^2 dx =: C_0.$$

The inequalities in the statement of the lemma follow immediately. □

Lemma 6.2 *Let $u(x,t)$ be a solution to initial-boundary value problem (6.2)-(6.3); then*

$$\|u_x(\cdot,T)\|^2_{L^2(\Omega)} \le C_1 + C_0 \left(\frac{\beta(M)}{\varepsilon}\right)^2$$

where C_0 is the constant constructed in Lemma 6.1, $C_1 = \|u_x(\cdot,0)\|^2_{L^2(\Omega)}$, $M = \sup\{|u(x,t)| : (x,t) \in \Pi_T\}$, and $\beta(M) = \sup\{\|\Phi''(u)\| : u \le M\}$.

Proof. Differentiate (6.2) with respect to x and set $v = u_x$; then $v_t + \nabla\Phi(u)_{xx} = Bv_{xx}$. Take the inner product of this equation with v to obtain

$$\langle v, Bv_{xx}\rangle = \langle v, v_t\rangle + \langle v, \nabla\Phi(u)_{xx}\rangle.$$

It follows from this equation (and $v = u_x$) that

$$\langle v_x, \nabla\Phi(u)_x\rangle = \frac{1}{2}\frac{\partial |v|^2}{\partial t} + \frac{\partial}{\partial x}[\langle v, \nabla\Phi(u)_x\rangle - \langle v, Bv_x\rangle] + \langle u_{xx}, Bu_{xx}\rangle.$$

Integrating over Π_T, we have

$$\begin{aligned} &\tfrac{1}{2}\textstyle\int_0^1 |u_x|^2\Big|_0^T dx + \iint_{\Pi_T} \frac{\partial\langle u_x, u_t\rangle}{\partial x} + \langle u_{xx}, Bu_{xx}\rangle\, dx\, dt \\ &= \textstyle\iint_{\Pi_T} \langle v_x, \nabla\Phi(u)_x\rangle\, dx\, dt. \end{aligned} \tag{6.5}$$

Note that $\langle u_x, u_t\rangle = -\langle u, u_{xt}\rangle + \frac{\partial}{\partial t}\langle u, u_x\rangle$. Using the boundary conditions in (6.3), we obtain $\langle u_x, u_t\rangle|_0^1 = \frac{\partial}{\partial t}\langle u, u_x\rangle|_0^1$. As a consequence we have

$$\iint_{\Pi_T} \frac{\partial}{\partial x}\langle u_x, u_t\rangle\, dx\, dt = \int_0^T \langle u_x, u_t\rangle|_0^1\, dt = \langle u, u_x\rangle|_0^1|_0^T = 0$$

where we again have used the boundary conditions in (6.3).

Equation (6.5) implies

$$\begin{aligned} &\tfrac{1}{2}\textstyle\int_0^1 |u_x(x,T)|^2 dx + \varepsilon \iint_{\Pi_T} |u_{xx}|^2 dx\, dt \\ &\le \tfrac{1}{2}\textstyle\int_0^1 |u_x(x,0)|^2 dx + \iint_{\Pi_T} \langle v_x, \nabla\Phi(u)_x\rangle\, dx\, dt \\ &= \tfrac{1}{2}\|u_x(\cdot,0)\|^2_{L^2(\Omega)} + \textstyle\iint_{\Pi_T} \langle u_{xx}, \nabla\Phi(u)_x\rangle\, dx\, dt. \end{aligned}$$

Using the identity $\langle a,b\rangle \le \frac{1}{2}(\varepsilon|a|^2 + \frac{1}{\varepsilon}|b|^2)$ with $a = u_{xx}$ and $b = \nabla\Phi(u)_x$, we obtain

$$\begin{aligned}&\textstyle\int_0^1 |u_x(x,T)|^2 dx + \varepsilon \iint_{\Pi_T} |u_{xx}|^2 dx\,dt\\ &\quad\le \|u_x(\cdot,0)\|^2_{L^2(\Omega)} + \frac{1}{\varepsilon}\iint_{\Pi_T} |\nabla\Phi(u)_x|^2 dx\,dt\end{aligned}$$

and consequently

$$\begin{aligned}\|u_x(\cdot,T)\|^2_{L^2(\Omega)} &= \textstyle\int_0^1 |u_x(x,T)|^2 dx\\ &\le C_1 + \frac{1}{\varepsilon}\iint_{\Pi_T}\langle|\nabla\Phi(u)_x|^2\rangle\,dx\,dt\\ &\le C_1 + \frac{1}{\varepsilon}\iint_{\Pi_T}\|\Phi''(u)\|^2|u_x|^2 dx\,dt\\ &\le C_1 + \frac{1}{\varepsilon}\alpha(M)\iint_{\Pi_T}|u_x|^2 dx\,dt\\ &\le C_1 + \frac{C_0}{\varepsilon^2}\alpha(M)\end{aligned}$$

where $\alpha(M) = \sup\{\|\Phi''(u)\|^2 : u \le M\}$. From the standard inequality for the norm of a product of operators, we have $\alpha(M) \le \beta^2(M)$. Thus,

$$\|u_x(\cdot,T)\|^2_{L^2(\Omega)} \le C_1 + C_0\left(\frac{\beta(M)}{\varepsilon}\right)^2$$

which completes the lemma. □

Theorem 6.3 *Assuming the growth condition (6.4), the initial-boundary value problem (6.2)-(6.3) has a global solution which tends uniformly to zero on* $\overline{\Omega}$ *as* $t \to \infty$.

Proof. We can write

$$\begin{aligned}|u(x,t)|^2 &= \langle u,u\rangle\\ &= \textstyle\int_0^x \frac{\partial}{\partial x}\langle u,u\rangle\,dx\\ &= 2\textstyle\int_0^x\langle u_x,u\rangle\,dx\\ &\le 2\|u_x(\cdot,t)\|_{L^2(\Omega)}\,\|u(\cdot,t)\|_{L^2(\Omega)}.\end{aligned}$$

By the bounds constructed in Lemmas 6.1 and 6.2, we have

$$|u(x,t)|^2 \le 2\sqrt{2C_0}\sqrt{C_1 + C_0\left(\frac{\beta(M)}{\varepsilon}\right)^2} \le K_1 + \frac{K_0}{\varepsilon}\beta(M) \tag{6.6}$$

for some positive constants K_0 and K_1.

The condition (6.4) implies that for each $\delta > 0$, there is an M_0 sufficiently large so that $[1+\beta(M)]/M^2 \le \delta$ for $M \ge M_0$. Choose δ so that $\delta K_0/\varepsilon < 1$; then taking the supremum of (6.6) over Π_T, we have

$$M^2 \le K_1 + \frac{\delta K_0}{\varepsilon}M^2$$

which, for our choice of δ, implies

$$|u(x,t)|^2 \le M^2 \le \frac{K_1}{1 - \frac{\delta K_0}{\varepsilon}}.$$

This gives a global *a priori* bound, $|u(x,t)| \le M_0$, where M_0 is independent of t. Thus, the solution $u(x,t)$ exists globally.

We now obtain the asymptotic behavior of the solution $u(x,t)$ as $t \to \infty$. By Lemma 6.1 we have $\|u_x\|^2_{L^2(\Pi_T)} \le C_0/\varepsilon$. This bound implies that for $N(t) = \|u_x(\cdot,t)\|^2_{L^2(\Omega)}$, $\int_0^\infty N(t)\,dx < \infty$. Using the identity

$$\begin{aligned} N'(t) &= 2\int_0^1 \langle u_x, u_{xt}\rangle\,dx \\ &= -2\int_0^1 \langle u_{xx}, u_t\rangle\,dx \end{aligned}$$

and using the dissipativity of B, one can show that

$$\begin{aligned} N'(t) &\le -\varepsilon\int_0^1 |u_{xx}|^2 dx + \tfrac{1}{\varepsilon}\int_0^1 \|\Phi''(u)\||u_x|^2 dx \\ &\le \tfrac{1}{\varepsilon}\|\Phi''(M_0)\|^2 \int_0^1 |u_x|^2 dx \end{aligned}$$

so that

$$\int_0^\infty |N'(t)|\,dx \le \frac{1}{\varepsilon}\|\Phi''(M_0)\|^2 \int_0^\infty N(t)\,dt < \infty.$$

Thus, $N(t)$ has finite total variation which implies $N(\infty)$ exists. Since $\int_0^\infty N(t)\,dt < \infty$, it must be that $N(\infty) = 0$. We have established

$$\|u_x(\cdot,t)\|^2_{L^2(\Omega)} \to 0 \text{ as } t \to \infty.$$

By Lemma 6.1 we have $\|u(\cdot,t)\|^2_{L^2(\Omega)} \le 2C_0$. As a result,

$$|u(x,t)|^2 \le 2\|u_x(\cdot,t)\|_{L^2(\Omega)}\,\|u(\cdot,t)\|_{L^2(\Omega)} \le 2\sqrt{2C_0}\|u_x(\cdot,t)\|_{L^2(\Omega)},$$

so $|u(x,t)| \to 0$ as $t \to \infty$ uniformly in x. □

The proof is valid for the boundary conditions $u(0,t) = u(1,t) = c$ where c is a nonzero constant since $\tilde{u} = u - c$ is also a solution to (6.2).

6.2 Induction Model for a Reactive-Euler System

In Section 1.4, we developed the reactive-Euler model (1.42) when $O(t_R) = t_A \ll t_C$. In one spatial dimension, this can be written as

$$\begin{aligned} &\phi_t - \tfrac{\gamma-1}{\gamma}\tilde{p}_t = \delta e^\phi \\ &\tilde{v}_t + \tfrac{1}{\gamma}\left(\tfrac{t_C\alpha}{t_A}\right)^2 \tilde{p}_x = 0 \\ &\tilde{v}_x + \tfrac{1}{\gamma}\tilde{p}_t = \delta e^\phi \end{aligned} \tag{6.7}$$

for $(x,t) \in \mathbb{R} \times (0,\infty)$ and where $\gamma \geq 1$ is the gas constant, h is the heat release, and $\delta = h/\gamma$ is the Frank- Kamenetski parameter. Initial data is given by

$$\phi(x,0) = \phi_0(x), \ \ \tilde{p}(x,0) = \tilde{p}_0(x), \ \ \tilde{v}(x,0) = \tilde{v}_0(x), \ \ x \in \mathbb{R}, \tag{6.8}$$

with all functions continuous and bounded.

Set

$$a = \frac{\gamma - 1}{\gamma}, \ \ b = \delta, \ \ c = \frac{1}{\gamma}\left(\frac{t_C \alpha}{t_A}\right)^2, \ \text{ and } \ d = \frac{1}{\gamma};$$

then with $w = \phi - a\tilde{p}$, (6.7)-(6.8) can be written as

$$\begin{aligned} w_t &= be^{w+a\tilde{p}} \\ \tilde{v}_t + c\tilde{p}_x &= 0 \\ \tilde{p}_t + \tfrac{1}{d}\tilde{v}_x &= \tfrac{b}{d}e^{w+a\tilde{p}} \end{aligned} \tag{6.9}$$

with

$$w(x,0) = \phi_0(x) - a\tilde{p}_0(x), \ \ \tilde{v}(x,0) = \tilde{v}_0(x), \ \ \tilde{p}(x,0) = \tilde{p}_0(x). \tag{6.10}$$

Using the change of coordinates

$$\begin{pmatrix} \overline{v} \\ \overline{p} \end{pmatrix} = \begin{pmatrix} 1 & 1 \\ -(cd)^{-1/2} & (cd)^{-1/2} \end{pmatrix}^{-1} \begin{pmatrix} \tilde{v} \\ \tilde{p} \end{pmatrix},$$

(6.9)-(6.10) is equivalent to

$$\begin{aligned} w_t &= be^{w+\mu(\overline{p}-\overline{v})} \\ \overline{v}_t - \lambda\overline{v}_x &= -\tfrac{b\lambda}{2}e^{w+\mu(\overline{p}-\overline{v})} \\ \overline{p}_t + \lambda\overline{p}_x &= \tfrac{b\lambda}{2}e^{w+\mu(\overline{p}-\overline{v})} \end{aligned} \tag{6.11}$$

with $\mu = a(cd)^{-1/2}$, $\lambda = (c/d)^{1/2}$, and initial conditions

$$\begin{aligned} w(x,0) &= \phi_0(x) - a\tilde{p}_0(x) \\ \overline{v}(x,0) &= \left(\tfrac{cd}{2}\right)^{1/2}\left[\tfrac{1}{(cd)^{1/2}}\tilde{v}_0(x) - \tilde{p}_0(x)\right] \\ \overline{p}(x,0) &= \left(\tfrac{cd}{2}\right)^{1/2}\left[\tfrac{1}{(cd)^{1/2}}\tilde{v}_0(x) + \tilde{p}_0(x)\right]. \end{aligned} \tag{6.12}$$

Setting

$$u = \mu\overline{p}, \ \ v = -\mu\overline{v}, \ \ A = b = \delta, \ \text{ and } \ B = \frac{b\lambda\mu}{2} = \frac{(\gamma-1)\delta}{2},$$

(6.11)-(6.12) is equivalent to

$$\begin{aligned} w_t &= Ae^{w+u+v} \\ u_t + \lambda u_x &= Be^{w+u+v} \\ v_t - \lambda v_x &= Be^{w+u+v} \end{aligned} \tag{6.13}$$

with initial conditions

$$\begin{aligned} w(x,0) &= \phi_0(x) - a\tilde{p}_0(x) =: \overline{w}(x) \\ u(x,0) &= \tfrac{a}{2}[(cd)^{-1/2}\tilde{v}_0(x) + \tilde{p}_0(x)] =: \overline{u}(x) \\ v(x,0) &= -\tfrac{a}{2}[(cd)^{-1/2}\tilde{v}_0(x) - \tilde{p}_0(x)] =: \overline{v}(x). \end{aligned} \tag{6.14}$$

We assume henceforth that $A + 2B = 1$.

Let $c^+ = \max\{A, B\}$, $c^- = \min\{A, B\}$, $m^+ = \max\{\overline{w}(x), \overline{u}(x), \overline{v}(x) : x \in \overline{\Omega}\}$, and $m^- = \min\{\overline{w}(x), \overline{u}(x), \overline{v}(x) : x \in \overline{\Omega}\}$, and consider

$$z' = c^{\pm}e^{3z}, \quad z(0) = m^{\pm}.$$

By comparison with (6.13)-(6.14),

$$\ln(e^{-3m^-} - 3c^-t)^{-1/3} \le \left\{ \begin{matrix} w(x,t) \\ u(x,t) \\ v(x,t) \end{matrix} \right\} \le \ln(e^{-3m^+} - 3c^+t)^{-1/3}$$

and hence every solution (w, u, v) of (6.13)-(6.14) blows up in finite time T with

$$\frac{1}{3c^+ \exp(3m^+)} \le T \le \frac{1}{3c^- \exp(3m^-)}.$$

Note that $\phi(x,t) = w(x,t) + u(x,t) + v(x,t)$. Assume that $\phi(x,t)$ blows up at x_m at time T. We would like to describe how the blowup singularity evolves at (x_m, T). Make the backward similarity change of variables

$$\tau = -\ln(T - \tau), \quad \eta = \frac{x - x_0}{(T-t)^{1/2}}$$

with

$$\begin{aligned} W &= w + A\ln(T-t), \\ U &= u + B\ln(T-t), \\ V &= v + B\ln(T-t), \\ \Phi &= \phi + \ln(T-t) = W + U + V; \end{aligned}$$

then (6.13) becomes

$$\begin{aligned} &W_\tau + \tfrac{\eta}{2} W_\eta = A(e^\Phi - 1) \\ &U_\tau + \tfrac{\eta}{2} U_\eta + \lambda e^{-\tau/2} U_\eta = B(e^\Phi - 1) \\ &V_\tau + \tfrac{\eta}{2} V_\eta - \lambda e^{-\tau/2} V_\eta = B(e^\Phi - 1) \\ &\Phi_\tau + \tfrac{\eta}{2} \Phi_\eta + \lambda e^{-\tau/2}(U_\eta - V_\eta) = e^\Phi - 1. \end{aligned} \tag{6.15}$$

To describe how the blowup evolves would require us to analyze the behavior of solutions of (6.15) as τ becomes infinite. To get an idea of what to expect or hope for, let us consider the much easier problem when there is no drift; that is, $\lambda = 0$. The temperature ϕ blows up at

$$T_\phi = e^{-\phi_0(x_m)}$$

where x_m is an absolute maximum point for ϕ_0. Then we know when and where blowup occurs. We can also describe precisely how the blowup singularity evolves at x_m.

Let $z = \phi + \ln(T - t)$; then z is the solution of

$$\begin{aligned} &z_\tau + \tfrac{\eta}{2} z_\eta = e^z - 1 \\ &z(\eta, -\ln T) = z_0(\eta) = \phi_0(\eta T^{1/2} + x_m) + \ln T \end{aligned}$$

which can be explicity solved to give

$$z(\eta, \tau) = -\ln\left[1 - e^\tau \left(1 - e^{-z_0(\eta e^{-\tau/2})}\right)\right].$$

Thus,

$$\lim_{\tau \to \infty} z(\eta, \tau) = -\ln\left(1 - z_0''(0) e^{-z_0(0)} \frac{\eta^2}{2}\right) = -\ln\left(1 - \frac{K\eta^2}{2}\right) =: \bar{z}(\eta).$$

From this, we conclude that when $\lambda = 0$,

$$\phi(x, t) + \ln[(T - t) - K(x - x_m)^2] \to 0$$

uniformly for $(x - x_m)^2 \leq \eta(T - t)$ as $t \to T^-$ which gives us a description of how the blowup singularity evolves. We would expect a similar type of behavior for (6.13)-(6.14).

6.3 The Full One-Dimensional Gas Model

In this section we consider a system of equations modeling the behavior of a heat-conductive viscous reactive compressible gas bounded by two parallel plates. Assume that its describing parameters vary spatially only in the direction perpendicular to the plates. In Euler coordinates, we have:

$$\begin{aligned} &\rho_t + (v\rho)_y = 0 \\ &\rho[v_t + vv_y] = \lambda_1 v_{yy} - k(\rho\theta)_y \\ &\rho[\theta_t + v\theta_y] = \lambda_2 \theta_{yy} + \lambda_1 v_y^2 - k\rho\theta v_y + \delta\rho f(\rho,\theta,z) \\ &\rho[z_t + vz_y] = \lambda_3(\rho z_y)_y - \rho f(\rho,\theta,z) \end{aligned} \tag{6.16}$$

where k, δ, and λ_i $(i = 1,2,3)$ are positive constants, where $t \geq 0$ is the time, and where $y \in [0,1] \subset \mathbb{R}$ is the one -dimensional space variable. The variables ρ, v, θ, and z represent the density, velocity, temperature, and concentration of unburned fuel, respectively. Let $\Omega = (0,1)$ and $\partial\Omega = \{0,1\}$. The initial conditions for (6.16) will be

$$\begin{aligned} \rho(y,0) = \rho_0(y), \quad v(y,0) = v_0(y) \\ \theta(y,0) = \theta_0(y), \quad z(y,0) = z_0(y) \end{aligned} \quad , \; y \in \Omega. \tag{6.17}$$

The results proved in this section involve the boundary conditions

$$\begin{aligned} &v(y,t) = 0, \quad \theta_y(y,t) = 0 \\ &z_y(y,t) = 0 \end{aligned} \quad , \; (y,t) \in \partial\Omega \times (0,\infty). \tag{6.18}$$

and represent a thermally insulated boundary. Similar results can be proved for the noninsulated boundary conditions

$$\begin{aligned} &v(y,t) = 0, \quad z_y(y,t) = 0 \\ &a[\theta(y,t) - T] - b\theta_y(y,t) = 0 \end{aligned} \quad , \; (y,t) \in \partial\Omega \times (0,\infty), \tag{6.19}$$

where $a > 0$, $b > 0$, $a + b > 0$, and $T > 0$.

In establishing global *a priori* bounds for global existence, we rewrite system (6.16) in terms of the Lagrange variable $x(y,t) = \int_0^y \rho(r,t)\,dr$. Treating the functions involved as functions of (x,t), system (6.16) is transformed into

$$\begin{aligned} &\rho_t = -\rho^2 v_x \\ &v_t = \lambda_1(\rho v_x)_x - k(\rho\theta)_x \\ &\theta_t = \lambda_2(\rho\theta_x)_x + \lambda_1 \rho v_x^2 - k\rho\theta v_x + \delta f(\rho,\theta,z) \\ &z_t = \lambda_3(\rho^2 z_x)_x - f(\rho,\theta,z) \end{aligned} \tag{6.20}$$

with initial conditions

$$\begin{aligned} \rho(x,0) = \rho_0(x), \quad v(x,0) = v_0(x) \\ \theta(x,0) = \theta_0(x), \quad z(x,0) = z_0(x) \end{aligned} \quad , \; x \in \Omega. \tag{6.21}$$

and boundary conditions

$$\begin{matrix} v(x,t) = 0, & \theta_x(x,t) = 0 \\ z_x(x,t) = 0 & \end{matrix} \quad , \ (x,t) \in \partial\Omega \times (0,\infty). \tag{6.22}$$

or

$$\begin{matrix} v(x,t) = 0, \ z_x(x,t) = 0 \\ a[\theta(x,t) - T] - b\theta_x(x,t) = 0 \end{matrix} \quad , \ (x,t) \in \partial\Omega \times (0,\infty), \tag{6.23}$$

In the remaining discussion we will use the following notation. For any function $g : \Omega \to \mathbb{R}$, define the constants

$$m_g = \inf_{x\in\Omega} g(x) \text{ and } M_g = \sup_{x\in\Omega} g(x).$$

For any function $h : \Omega \times [0,T] \to \mathbb{R}$, define the functions

$$m_h(t) = \inf_{x\in\Omega} h(x,t) \text{ and } M_h(t) = \sup_{x\in\Omega} h(x,t).$$

In addition for the initial-boundary value problem (6.20)-(6.21)-(6.22), we make the assumptions:

$$\begin{aligned} & 0 < m_{\rho_0} \le \rho_0(x) \le M_{\rho_0} < \infty, \\ & -\infty < m_{v_0} \le v_0(x) \le M_{v_0} < \infty, \\ & 0 < m_{\theta_0} \le \theta_0(x) \le M_{\theta_0} < \infty, \\ & 0 \le m_{z_0} \le z_0(x) \le M_{z_0} < \infty, \end{aligned} \tag{6.24}$$

$$\int_0^1 \frac{1}{\rho_0(x)}\,dx = 1, \tag{6.25}$$

and $f : [0,\infty)^3 \to [0,\infty)$ is continuous and globally Lipschitz on $[0,\overline{\rho}] \times [0,\infty) \times [0,\overline{z}]$ for all $\overline{\rho} > 0$ and for all $\overline{z} > 0$.

Definition 6.1 *Consider the set of functions (ρ, v, θ, z) satisfying*

$$\begin{aligned} & \rho \in L^\infty([0,T], W_2^1(\Omega)), \ \rho_t \in L^\infty([0,T], L^2(\Omega)), \ \textit{and} \\ & (v,\theta,z) \in L^\infty([0,T], W_2^1(\Omega)) \cap L^2([0,T], W_2^2(\Omega)) \cap W_2^1([0,T], L^2(\Omega)). \end{aligned}$$

A generalized solution (ρ, v, θ, z) is a function satisfying equations (6.20) almost everywhere and assumes the initial-boundary conditions in the sense of traces. For $1 \le p \le \infty$ we denote by W_p^k the Sobolev space of all functions whose derivatives up to order k are in L^p.

Local existence for initial-boundary value problem (6.20)-(6.21)-(6.22) can be found in [KZH2]. The key theorem for global existence is the following theorem which establishes global *a priori* bounds for the solution.

Theorem 6.4 *If* $(\rho_0(x), v_0(x), \theta_0(x), z_0(x)) \in W_2^1(\Omega)$ *and if* (ρ, v, θ, z) *is any generalized solution of (6.20)-(6.21)-(6.22), then for any* $T > 0$ *there exists a constant* $C > 0$ *such that*

$$|v(x,t)| \leq C,\ C^{-1} \leq \rho(x,t) \leq C,$$
$$C^{-1} \leq \theta(x,t) \leq C, \quad \text{and } 0 \leq z(x,t) \leq C$$

for all $(x,t) \in \overline{\Omega} \times [0,T] =: R_T$.

The proof of Theorem 6.4 will consist of a sequence of lemmas which establish the bounds on the functions. As a consequence of the estimates, we can prove

Theorem 6.5 *If* $(\rho_0(x), v_0(x), \theta_0(x), z_0(x)) \in W_2^1(\Omega)$, *then IBVP (6.20)-(6.21)-(6.22) has a unique generalized solution for* $(x,t) \in \overline{\Omega} \times [0,\infty)$.

By imposing additional smoothness on the initial data, we can prove the next theorem using the results in [KZH1],[NAS].

Theorem 6.6 *If* $(v_0, \theta_0, z_0) \in C^{2+\alpha}(\Omega)$ *and* $\rho_0 \in C^{1+\alpha}(\Omega)$ *for* $0 < \alpha < 1$, *then initial-boundary value problem (6.20)-(6.21)-(6.22) has a unique classical solution* $(\rho, v, \theta, z)(x,t)$ *on* $\overline{\Omega} \times [0,\infty)$ *with*

$$(v, \theta, z) \in C^{2+\alpha, 1+\alpha/2}(\overline{\Omega} \times [0,\infty))$$

and

$$\rho \in C^{1+\alpha, 1+\alpha/2}(\overline{\Omega} \times [0,\infty)).$$

The next result is not proved here; these results can be found in [BRE] and are similar to those used in this chapter.

Theorem 6.7 *Let* $\rho_0 \in C^{1+\alpha}(\Omega)$ *and* $(v_0, \theta_0, z_0) \in C^{2+\alpha}(\Omega)$ *for* $\alpha \in (0,1)$. *Suppose that*

$$v_0(x) = z_0'(x),\ \lambda_1 v_0(x) - k[\rho_0(x)\theta_0(x)]' = 0,$$

and

$$a[\theta_0(x) - T] - b\theta_0'(x) = 0$$

for $x \in \partial\Omega$. *In the event that* $b = 0$, *also assume that*

$$\rho_0 v_0 \theta_0' - \lambda_2 \theta_0'' - \lambda_1 (v_0')^2 + k\theta_0 v_0 - \delta \rho_0 f(\rho_0, \theta_0, z_0) = 0$$

for $x \in \partial\Omega$; *then initial-boundary value problem (6.20)- (6.21)-(6.23) has a unique classical solution on* $\overline{\Omega} \times [0,\infty)$.

These results show that even when heat is added at the boundary and the gas is reactive, for arbitrarily large Lipschitz continuous initial data, *no shocks develop.*

6.4 Energy Estimates and Density Bounds

From (6.20a) we obtain $\frac{\partial}{\partial t}\left(\frac{1}{\rho}\right) = v_x$. Integrating with respect to t gives us

$$\rho(x,t) = \left[\frac{1}{\rho_0(x)} + \int_0^t v_x(x,\tau)\,d\tau\right]^{-1}$$

which is positive for t sufficiently small since $\rho_0^{-1}(x) > 0$ on Ω. In addition, integrating the same equation with respect to x and using the boundary conditions for v gives us $(\int_0^1 \rho^{-1}\,dx)_t \equiv 0$. Thus,

$$\int_0^1 \rho^{-1}\,dx = \int_0^1 \rho_0^{-1}\,dx = 1.$$

Although $\rho > 0$ for small time, we will eventually obtain an upper bound for ρ and bounds on $|v_x|$. By bootstrapping, we find that $\rho > 0$ for larger times. Using $\rho > 0$, we can use standard comparison results on the equations in (6.20) to obtain the following bounds on the temperature and concentration functions.

Lemma 6.8 *The functions $\theta(x,t)$ and $z(x,t)$ satisfy the conditions*

$$\theta(x,t) \geq 0 \quad \textit{and} \quad 0 \leq z(x,t) \leq M_{z_0}$$

for all $(x,t) \in R_T$.

Proof. From (6.20c) we have

$$\theta_t = \lambda_2(\rho\theta_x)_x + \lambda_1\rho v_x^2 - k\rho v_x\theta + \delta f(\rho,\theta,z) \geq \lambda_2(\rho\theta_x)_x - (k\rho v_x)\theta$$

on R_T since $\rho > 0$, $\lambda_1 > 0$, $\delta > 0$, and $f(\rho,\theta,z) \geq 0$. On the parabolic boundary of R_T we have $\theta_0(x) > 0$ on $\overline{\Omega}$ and $\theta_x(\partial\Omega,t) = 0$, so by the maximum principle, $\theta(x,t) \geq 0$ on R_T.

From (6.20d) we have

$$z_t = \lambda_3(\rho^2 z_x)_x - f(\rho,\theta,z) \leq \lambda_3(\rho^2 z_x)_x$$

on R_T since $f(\rho,\theta,z) \geq 0$. On the parabolic boundary of R_T we have $z_0(x) \leq M_{z_0}$ on $\overline{\Omega}$ and $z_x(\partial\Omega,t) = 0$, so by the maximum principle, $z(x,t) \leq M_{z_0}$ on R_T.

Using the full equation for z and using Lipschitz continuity of f in the z-component, since $z_0 \geq 0$ on $\overline{\Omega}$ and $z_x(\partial\Omega,t) = 0$ we have by the maximum principle $z(x,t) \geq 0$ on R_T. □

We now consider the energy densities

$$w(x,t) = \theta(x,t) + \frac{1}{2}v^2(x,t)$$

and

$$e(x,t) = \theta(x,t) + \frac{1}{2}v^2(x,t) + \delta z(x,t).$$

The total energy $\int_0^1 w(x,t)\,dx$ may increase in time due to the heat generated by the chemical reaction. However, using $e(x,t)$ we can prove the following result.

Lemma 6.9 *The function $e(x,t)$ satisfies the condition*

$$\int_0^1 e(x,t)\,dx = A_1$$

where A_1 is a positive constant. As a result we have the bounds

$$\int_0^1 \theta(x,t)\,dx \le A_1 \quad \textit{and} \quad \int_0^1 v^2(x,t)\,dx \le 2A_1.$$

Proof. Using the equations in (6.20) we have

$$\begin{aligned}
\tfrac{d}{dt}\textstyle\int_0^1 e(x,t)\,dx &= \tfrac{d}{dt}\textstyle\int_0^1 [\theta + \frac{1}{2}v^2 + \delta z]\,dx \\
&= \textstyle\int_0^1 [\theta_t + v v_t + \delta z_t]\,dx \\
&= \textstyle\int_0^1 \{[\lambda_2(\rho\theta_x)_x + \lambda_1 \rho v_x^2 - k\rho\theta v_x + \delta f(\rho,\theta,z)] \\
&\quad + [\lambda_1 v(\rho v_x)_x - kv(\rho\theta)_x] \\
&\quad + [\delta\lambda_3(\rho^2 z_x)_x - \delta f(\rho,\theta,z)]\}\,dx \\
&= \lambda_1 \textstyle\int_0^1 [\rho v_x^2 + v(\rho v_x)_x]\,dx + \lambda_2 \int_0^1 (\rho\theta_x)_x\,dx \\
&\quad + \delta\lambda_3 \textstyle\int_0^1 (\rho^2 z_x)_x\,dx - k\int_0^1 [\rho\theta v_x + v(\rho\theta)_x]\,dx \\
&= \lambda_1 \textstyle\int_0^1 (\rho v v_x)_x\,dx + \lambda_2 \int_0^1 (\rho\theta_x)_x\,dx \\
&\quad + \delta\lambda_3 \textstyle\int_0^1 (\rho^2 z_x)_x\,dx - k\int_0^1 (\rho v\theta)_x\,dx \\
&= [\lambda_1 \rho v v_x + \lambda_2 \rho\theta_x + \delta\lambda_3 \rho^2 z_x - k\rho v\theta]|_{x=0}^{x=1} \\
&= 0
\end{aligned}$$

where the boundary conditions (6.22) have been used. Thus,

$$\begin{aligned}
\textstyle\int_0^1 e(x,t)\,dx &= \textstyle\int_0^1 e(x,0)\,dx \\
&= \textstyle\int_0^1 [\theta_0(x) + \frac{1}{2}v_0^2(x) + \delta z_0(x)]\,dx \\
&=: A_1.
\end{aligned}$$

By Lemma 6.8 we had $z(x,t) \ge 0$. Thus,

$$\int_0^1 \theta(x,t)\,dx + \frac{1}{2}\int_0^1 v^2(x,t)\,dx \le A_1$$

which gives the stated bounds. □

For each $t \in [0,T]$, there is a value $a(t)$ such that $\rho(a(t),t) = 1$. The existence of this value is assured by $\int_0^1 \rho^{-1}\,dx = 1$ and the fact that $\rho(\cdot,t)$ has a point of continuity $x_0 \in \Omega$.

From (6.20a) we have $\rho_t = -\rho^2 v_x$ which implies $\rho v_x = -(\ln\rho)_t$. Using this in (6.20b) we obtain

$$v_t = -\lambda_1(\ln\rho)_{tx} - k(\rho\theta)_x = -\lambda_1(\ln\rho)_{tx} - p_x$$

where the pressure $p(x,t) = k\rho(x,t)\theta(x,t)$. Integrating with respect to t (with lower limit 0) we get

$$\frac{\partial}{\partial x}\left[\lambda_1 \ln\rho(x,t) - \lambda_1 \ln\rho_0(x) + \int_0^t p(x,\tau)\,d\tau\right] = v_0(x) - v(x,t).$$

Integrating with respect to x (with lower limit $a(t)$) we get

$$\begin{aligned}&\int_{a(t)}^x [v_0(\xi) - v(\xi,t)]\,d\xi\\ &\quad = \lambda_1 \ln\left[\frac{\rho(x,t)\rho_0(a(t))}{\rho_0(x)}\right] + \int_0^t p(x,\tau)\,d\tau - \int_0^t p(a(t),\tau)\,d\tau\end{aligned}$$

where we have used $\rho(a(t),t) = 1$. Exponentiating gives us

$$\begin{aligned}&\rho(x,t)e^{\frac{1}{\lambda_1}\int_0^t p(x,\tau)d\tau}\\ &\quad = \rho_0(x)\rho_0^{-1}(a(t))e^{\frac{1}{\lambda_1}\int_0^t p(a(t),\tau)\,d\tau}e^{\frac{1}{\lambda_1}\int_{a(t)}^x [v_0(\xi)-v(\xi,t)]\,d\xi}\end{aligned}$$

and multiplying both sides by $k\theta$ (and using $p = k\rho\theta$), we obtain

$$\begin{aligned}\frac{\partial}{\partial t}\left[\lambda_1 \exp\left(\frac{1}{\lambda_1}\int_0^t p(x,\tau)\,d\tau\right)\right] &= p(x,t)\exp\left(\frac{1}{\lambda_1}\int_0^t p(x,\tau)\,d\tau\right)\\ &= k\theta(x,t)\rho_0(x)Y(t)B(x,t)\end{aligned} \qquad (6.26)$$

where

$$Y(t) := \rho_0^{-1}(a(t))\exp\left(\frac{1}{\lambda_1}\int_0^t p(a(t),\tau)\,d\tau\right)$$

and

$$B(x,t) := \exp\left(\frac{1}{\lambda_1}\int_{a(t)}^x [v_0(\xi) - v(\xi,t)]\,d\xi\right).$$

Lemma 6.10 *There are positive constants A_2 and A_3 such that*

$$\frac{1}{A_2} \le B(x,t) \le A_2 \quad \textit{and} \quad \frac{1}{A_3} \le Y(t) \le A_3$$

for all $(x,t) \in R_T$.

Proof. Note that

$$\begin{aligned}\left|\int_{a(t)}^{x} v(\xi,t)\,d\xi\right| &\le \int_0^1 |v(\xi,t)|\,d\xi \\ &\le \left(\int_0^1 v^2(\xi,t)\,d\xi\right)^{1/2}\end{aligned}$$

where the last inequality comes from Hölder's inequality. From Lemma 6.9 we had $\int_0^1 v^2(\xi,t)\,d\xi \le 2A_1$, so $\left|\int_{a(t)}^{x} v(\xi,t)\,d\xi\right| \le \sqrt{2A_1}$. Thus,

$$\begin{aligned}\frac{1}{\lambda_1}\int_{a(t)}^{x}[v_0(\xi)-v(\xi,t)]\,d\xi &\le \frac{1}{\lambda_1}\left|\int_{a(t)}^{x}[v_0(\xi)-v(\xi,t)]\,d\xi\right| \\ &\le \frac{1}{\lambda_1}\left(\int_0^1 |v_0(\xi)|\,d\xi + \sqrt{\int_0^1 v^2(\xi,t)\,d\xi}\right) \\ &= \frac{1}{\lambda_1}\left(\|v_0\|_{L^1(\Omega)} + \sqrt{2A_1}\,\right).\end{aligned}$$

Since $\exp(-|Q|) \le \exp(Q) \le \exp(|Q|)$, we have

$$\begin{aligned}A_2^{-1} &= \exp\left(-\lambda_1^{-1}\left(\|v_0\| + \sqrt{2A_1}\,\right)\right) \\ &\le B(x,t) \\ &\le \exp\left(\lambda_1^{-1}\left(\|v_0\| + \sqrt{2A_1}\,\right)\right) = A_2.\end{aligned}$$

Integrate equation (6.26) with respect to t to obtain

$$\lambda_1\left[\exp\left(\frac{1}{\lambda_1}\int_0^t p(x,\tau)\,d\tau\right) - 1\right] = k\rho_0(x)\int_0^t \theta(x,\tau)Y(\tau)B(x,\tau)\,d\tau$$

or

$$\exp\left(\frac{1}{\lambda_1}\int_0^t p(x,\tau)\,d\tau\right) = 1 + \frac{k\rho_0(x)}{\lambda_1}\int_0^t \theta(x,\tau)Y(\tau)B(x,\tau)\,d\tau.$$

Multiplying by $p(x,t)$ and using (6.26) gives us

$$p(x,t)\left[1 + \frac{k\rho_0(x)}{\lambda_1}\int_0^t \theta(x,\tau)Y(\tau)B(x,\tau)\,d\tau\right] = k\rho_0(x)\theta(x,t)Y(t)B(x,t)$$

or, using $p = k\rho\theta$,

$$\rho(x,t) = \frac{\rho_0(x)Y(t)B(x,t)}{1 + \frac{k\rho_0(x)}{\lambda_1}\int_0^t Y(\tau)B(x,\tau)\theta(x,\tau)\,d\tau}. \qquad (6.27)$$

Equation (6.27) implies

$$\rho^{-1}(x,t)Y(t) = B^{-1}(x,t)\left[\rho_0^{-1}(x) + \frac{k}{\lambda_1}\int_0^t Y(\tau)B(x,\tau)\theta(x,\tau)\,d\tau\right].$$

Integrating both sides with respect to x and using $\int_0^1 \rho^{-1}\,dx = 1$, we have

$$Y(t) = \int_0^1 B^{-1}(x,t)\left[\rho_0^{-1}(x) + \frac{k}{\lambda_1}\int_0^t Y(\tau)B(x,\tau)\theta(x,\tau)\,d\tau\right]dx.$$

Using $B(x,t) \le A_2$ and $B^{-1}(x,t) \le A_2^{-1}$ gives us the inequality

$$\begin{aligned}
Y(t) &\le A_2 \int_0^1 \rho_0^{-1}(x)\,dx + \tfrac{kA_2^2}{\lambda_1}\int_0^1\int_0^t Y(\tau)\theta(x,\tau)\,d\tau\,dx \\
&= A_2 + \tfrac{kA_2^2}{\lambda_1}\int_0^t \left[\int_0^1 \theta(x,\tau)\,d\tau\right] Y(\tau)\,d\tau \quad \text{(using (6.25))} \\
&\le A_2 + \tfrac{kA_1A_2^2}{\lambda_1}\int_0^t Y(\tau)\,d\tau \quad \text{(by Lemma 6.9)}.
\end{aligned}$$

By Gronwall's inequality,

$$\begin{aligned}
Y(t) &\le A_2 \exp\left(\tfrac{kA_1A_2^2}{\lambda_1}t\right) \\
&\le A_2 \exp\left(\tfrac{kA_1A_2^2}{\lambda_1}T\right) \\
&=: A_3
\end{aligned}$$

for $t \in [0,T]$. Since $\rho_0 \ge m_{\rho_0} > 0$ and $p \ge 0$, $Y(t)$ is bounded away from 0. Choose A_3 large so that $\frac{1}{A_3} \le Y(t) \le A_3$. □

Corollary 6.11 *There is a positive constant A_4 such that*

$$\rho(x,t) \le A_4$$

for all $(x,t) \in R_T$.

Proof. In equation (6.27), $1 + (k\rho_0/\lambda_1)\int_0^t YB\theta\,d\tau \ge 1$, $\rho_0(x) \le M_{\rho_0}$, $B(x,t) \le A_2$, and $Y(t) \le A_3$ imply $\rho(x,t) \le M_{\rho_0}A_2A_3 =: A_4$. □

In the next result we obtain an inequality for the temperature function $\theta(x,t)$. Using (6.27) and Lemma 6.10, we have

$$m_\rho(t) \ge \frac{1}{A_2A_3}\left[m_{\rho_0}^{-1} + \frac{kA_2A_3}{\lambda_1}\int_0^t \theta(x,\tau)\,d\tau\right]^{-1} > 0. \tag{6.28}$$

Lemma 6.12 *For any $\eta > 0$, there are positive constants $A_5 = A_5(\eta)$ and $A_6 = A_6(\eta)$ such that*

$$\theta^2(x,t) \le \eta J_1(t) + A_5 J_2(t) + A_6$$

for $(x,t) \in R_T$ where $J_1(t) = \int_0^1 \rho(x,t)\theta_x^2(x,t)\,dx$ and $J_2 = \int_0^t J_1(\tau)\,d\tau$.

Proof. Set $\psi(x,t) = \theta(x,t) - \int_0^1 \theta(\xi,t)\,d\xi$; then $\int_0^1 \psi(x,t)\,dx = 0$. For each $t \in [0,T]$, there must be an $x_1(t) \in [0,1]$ such that $\psi(x_1(t),t) = 0$ since $\int_0^1 \psi\,dx = 0$ and since $\psi(\cdot,t)$ has a point of continuity $x_0 \in \Omega$. Consequently,

$$\begin{aligned}
|\psi(x,t)|^{3/2} &= \tfrac{3}{2}\int_{x_1(t)}^{x} |\psi(x,t)|^{1/2}\operatorname{sgn}(\psi(\xi,t))\psi_\xi(\xi,t)d\xi \\
&= \tfrac{3}{2}\int_{x_1(t)}^{x} \left(\rho^{-1/2}|\psi|^{1/2}\right)\left(\rho^{1/2}[\operatorname{sgn}(\psi)]\psi_\xi\right)\,d\xi \\
&\le \tfrac{3}{2}\int_0^1 \left(\rho^{-1/2}|\psi|^{1/2}\right)\left(\rho^{1/2}[\operatorname{sgn}(\psi)]\psi_\xi\right)\,d\xi\,.
\end{aligned}$$

Using Hölder's inequality, we have

$$|\psi(x,t)|^{3/2} \le \frac{3}{2}\sqrt{\int_0^1 \rho^{-1}|\psi|\,d\xi}\,\sqrt{\int_0^1 \rho\psi_x^2\,d\xi}\,. \tag{6.29}$$

Since $\psi_x = \theta_x$, $\rho^{-1}(x,t) \le m_\rho^{-1}(t)$, and

$$\begin{aligned}
\int_0^1 |\psi(x,t)|\,dx &= \int_0^1 \left|\theta(x,t) - \int_0^1 \theta(\xi,t)\,d\xi\right|\,dx \\
&\le 2\int_0^1 |\theta(x,t)|\,dx \\
&\le 2A_1,
\end{aligned}$$

inequality (6.29) can be modified to

$$\begin{aligned}
|\psi(x,t)|^{3/2} &\le \tfrac{3}{2}m_\rho^{-1/2}(t)(2A_1)^{1/2}J_1^{1/2}(t) \\
&= \left(\tfrac{9A_1}{2}\right)^{1/2} m_\rho^{-1/2}(t)J_1^{1/2}(t).
\end{aligned}$$

Raising both sides to the $\frac{4}{3}$ power and using (6.28), we have

$$\begin{aligned}
\psi^2(x,t) &\le \left(\tfrac{9A_1}{2}\right)^{2/3} m_\rho^{-2/3}(t)J_1^{-2/3}(t) \\
&\le \left(\tfrac{9A_1}{2}\right)^{2/3}(A_2A_3)^{2/3}\left[m_{\rho_0}^{-1} + \tfrac{kA_2A_3}{\lambda_1}\int_0^t \theta(x,\tau)\,d\tau\right]^{2/3} J_1^{2/3}(t)
\end{aligned}$$

so that

$$\psi^2(x,t) \le \alpha_1\left[1 + \beta_1\int_0^1 \theta(x,\tau)\,d\tau\right]^{2/3} J_1^{2/3}(t) \tag{6.30}$$

where

$$\alpha_1 = \left(\frac{9A_1A2A_3}{2m_{\rho_0}}\right)^{2/3} \quad\text{and}\quad \beta_1 = \frac{kA_2A_3}{m_{\rho_0}\lambda_1}.$$

Since $\theta(x,t) = \psi(x,t) + \int_0^1 \theta(\xi,t)\,d\xi \le \psi(x,t) + A_1$, we have $\theta^2 \le (\psi + A_1)^2 \le 2(\psi^2 + A_1^2)$ where the last inequality follows from Cauchy's inequality ($2ab \le a^2 + b^2$). Using this in equation (6.30) yields

$$\theta^2(x,t) \le 2A_1^2 + 2\alpha_1\left[1 + \beta_1\int_0^t \theta(x,\tau)\,d\tau\right]^{2/3} J_1^{2/3}(t). \tag{6.31}$$

For any $\eta > 0$, choose $p = 3$, $q = 3/2$,

$$a = 2\alpha_1 \left[1 + \beta_1 \int_0^1 \theta(x,\tau)\, d\tau\right]^{2/3} \left(\frac{3\eta}{2}\right)^{-2/3},$$

and

$$b = \left(\frac{3\eta}{2}\right)^{2/3} J_1^{2/3}(t),$$

and apply Young's inequality ($ab \le \frac{a^p}{p} + \frac{b^q}{q}$ for $\frac{1}{p} + \frac{1}{q} = 1$) to inequality (6.31) to obtain

$$\begin{aligned} \theta^2(x,t) &\le 2A_1^2 + \tfrac{32\alpha_1^2}{27\eta^2}\left[1 + \beta \int_0^t \theta(x,\tau)\, d\tau\right]^2 + \eta J_1(t) \\ &\le 2A_1^2 + \tfrac{64\alpha_1^2}{27\eta^2}\left[1 + \beta_1^2 \left(\int_0^t \theta(x,\tau)\, d\tau\right)^2\right] + \eta J_1(t) \\ &= \alpha_2 + \beta_2 \left(\int_0^t \theta(x,\tau)\, d\tau\right)^2 + \eta J_1(t) \end{aligned}$$

where we have used the identity: $(1+a)^2 \le 2(1+a^2)$. By Hölder's inequality we have $\left(\int_0^t \theta\, d\tau\right)^2 \le t \int_0^t \theta^2 d\tau$ so

$$\theta^2(x,t) \le \alpha_2 + \beta_2 t \int_0^t \theta^2(x,\tau)\, d\tau \; + \eta J_1(t). \tag{6.32}$$

Set $I(t) := \int_0^t \theta^2(x,\tau)\, d\tau$ and $\beta_3 = \beta_2 T$; then $I(0) = 0$ and (6.32) can be rewritten as

$$I'(t) \le \alpha_2 + \beta_3 I(t) + \eta J_1(t) \;\text{ or }\; \left[e^{-\beta_3 t} I(t)\right]' \le [\alpha_2 + \eta J_1(t)] e^{-\beta_3 t}.$$

Thus, for $t \in [0,T]$,

$$I(t) \le e^{\beta_3 t} \int_0^t e^{-\beta_3 s}(\alpha_2 + \eta J_1(s))\, ds.$$

Replacing this in equation (6.32) yields

$$\begin{aligned} \theta^2(x,t) &\le \alpha_2 + \eta J_1(t) + \beta_3 e^{\beta_3 t} \int_0^t e^{-\beta_3 s}(\alpha_2 + \eta J_1(s))\, ds \\ &= \alpha_2(1 + \beta_3 e^{\beta_3 t} \int_0^t e^{-\beta_2 s} ds\,) + \eta J_1(t) + \beta_3 \int_0^t e^{\beta_3 (t-s)} J_1(s)\, ds \\ &\le A_6(\eta,T) + \eta J_1(t) + A_5(\eta,T) \int_0^t J_1(s)\, ds \\ &= \eta J_1(t) + A_5 J_2(t) + A_6 \end{aligned}$$

for $t \in [0,T]$ and where the constants are given by $A_5 = \beta_3 e^{\beta_3 T}$ and $A_6 = \alpha_2(1 + \beta_3 T e^{\beta_3 T})$. □

In order to get a positive lower bound on the density ρ, we consider the energy density $w(x,t) = \theta(x,t) + \frac{1}{2}v^2(x,t)$.

Lemma 6.13 *There is a positive constant A_7 such that*

$$\int_0^1 w^2(x,t)\,dx + \int_0^1 v^4(x,t)\,dx + J_2(t) \le A_7$$

for all $t \in [0,T]$. As a result we have the bounds

$$J_2(t) \le A_7 \quad \text{and} \quad \int_0^1 w^2\,dx \le A_7.$$

Proof. Using (6.20a,b), the time derivative of the energy density is

$$\begin{aligned} w_t &= (\theta + \tfrac{1}{2}v^2)_t \\ &= \theta_t + v v_t \\ &= [\lambda_2(\rho\theta_x)_x + \lambda_1 \rho v_x^2 - k\rho\theta v_x + \delta f(\rho,\theta,z)] \\ &\quad + v[\lambda_1(\rho v_x)_x - k(\rho\theta)_x] \\ &= (\lambda_2\rho\theta_x + \lambda_1\rho v v_x - k\rho\theta v)_x + \delta f(\rho,\theta,z). \end{aligned}$$

Therefore,

$$\begin{aligned} \tfrac{1}{2}\tfrac{d}{dt}\textstyle\int_0^1 w^2 dx &= \textstyle\int_0^1 w w_t\,dx \\ &= \textstyle\int_0^1 (\theta + \tfrac{1}{2}v^2)(\lambda_2\rho\theta_x + \lambda_1\rho v v_x - k\rho\theta v)_x\,dx \\ &\quad + \delta \textstyle\int_0^1 w f(\rho,\theta,z)\,dx \qquad (6.33) \\ &= -\textstyle\int_0^1 (\theta + \tfrac{1}{2}v^2)_x(\lambda_2\rho\theta_x + \lambda_1\rho v v_x - k\rho\theta v)\,dx \\ &\quad + \delta \textstyle\int_0^1 w f(\rho,\theta,z)\,dx \end{aligned}$$

where we have used integration by parts and the boundary conditions (6.22).

The global Lipschitz continuity of f and the bounds on z and ρ imply that

$$f(\rho,\theta,z) \le K_1\theta + f(\rho,0,z) \le K_1\theta + K_2 \le K_3(\theta+1)$$

where K_1 is the Lipschitz constant for f, K_2 is the bound on $f(\rho,0,z)$ for $(\rho,z) \in [0,A_4] \times [0,M_{z_0}]$, and where $K_3 = \max\{K_1,K_2\}$. Also, $w(\theta+1) = w(w+1-\frac{1}{2}v^2) \le w(w+1) = w^2 + w$, so we have

$$\begin{aligned} \delta \textstyle\int_0^1 w f(\rho,\theta,z)\,dx &\le \delta K_3 \textstyle\int_0^1 (w^2 + w)\,dx \\ &= \delta K_3 \left[\textstyle\int_0^1 w^2 dx + \int_0^1 (\theta + \tfrac{1}{2}v^2)\,dx\right] \\ &\le \delta K_3 \left(\textstyle\int_0^1 w^2 dx + A_1\right) \\ &\le K_4\left(1 + \textstyle\int_0^1 w^2 dx\right) \end{aligned}$$

where we have used Lemma 6.9 ($\|w(\cdot,t)\|_{L^1(\Omega)} \leq \|e(\cdot,t)\|_{L^1(\Omega)} = A_1$). Replacing this in (6.33) gives us

$$\begin{aligned}\int_0^1 ww_t\,dx &\leq -\int_0^1(\theta+\tfrac{1}{2}v^2)_x(\lambda_2\rho\theta_x+\lambda_1\rho v v_x - k\rho\theta v)\,dx\\ &\quad +K_4\left(1+\int_0^1 w^2dx\right)\\ &= -\lambda_2\int_0^1\rho\theta_x^2dx - \lambda_1\int_0^1\rho v^2v_x^2dx\\ &\quad -(\lambda_1+\lambda_2)\int_0^1\rho v v_x\theta_x\,dx\\ &\quad -k\left(\int_0^1\rho v\theta\theta_x\,dx + \int_0^1\rho\theta v^2 v_x\,dx\right)\\ &\quad +K_4\left(1+\int_0^1 w^2dx\right).\end{aligned} \tag{6.34}$$

Let $\alpha > 0$, $\beta > 0$, and $\gamma > 0$. By Cauchy's inequality,

$$-v\theta\theta_x \leq \frac{\alpha^2}{2}v^2\theta^2 + \frac{1}{2\alpha^2}\theta_x^2,\quad -\theta v^2 v_x \leq \frac{\beta^2}{2}v^2\theta^2 + \frac{1}{2\beta^2}v^2v_x^2,$$

and

$$-vv_x\theta_x \leq \frac{\gamma^2}{2}v^2v_x^2 + \frac{1}{2\gamma^2}\theta_x^2.$$

Using these in (6.34) yields

$$\begin{aligned}\int_0^1 ww_t\,dx &\leq K_5\int_0^1\rho\theta_x^2dx + K_6\int_0^1\rho v^2v_x^2dx\\ &\quad +K_7\int_0^1\rho v^2\theta^2dx + K_4\left(1+\int_0^1 w^2dx\right)\end{aligned}$$

where

$$\begin{aligned}K_5 &= \tfrac{\lambda_1+\lambda_2}{2\gamma^2} - \lambda_2 + \tfrac{k}{2\alpha^2},\\ K_6 &= \tfrac{\gamma^2(\lambda_1+\lambda_2)}{2} - \lambda_1 + \tfrac{k}{2\beta^2},\quad \text{and}\\ K_7 &= \tfrac{k}{2}(\alpha^2+\beta^2).\end{aligned}$$

The parameters α, β, and γ can be chosen so that $K_5 = -\lambda_2/2$ and $K_6 > 0$. Thus,

$$\begin{aligned}&\int_0^1 ww_t\,dx + \tfrac{\lambda_2}{2}\int_0^1\rho\theta_x^2dx\\ &\leq K_6\int_0^1\rho v^2v_x^2dx + K_7\int_0^1\rho v^2\theta^2dx + K_4\left(1+\int_0^1 w^2dx\right).\end{aligned} \tag{6.35}$$

Now consider the time derivative of $\int_0^1 v^4dx$. Using (6.20b), we have

$$\begin{aligned}\tfrac{1}{4}\tfrac{d}{dt}\int_0^1 v^4dx &= \int_0^1 v^3v_t\,dx\\ &= \int_0^1 v^3(\lambda_1\rho v_x - k\rho\theta)_x\,dx\\ &= -\int_0^1(v^3)_x(\lambda_1\rho v_x - k\rho\theta)\,dx\\ &= -3\lambda_1\int_0^1\rho v^2v_x^2dx + 3\int_0^1 k\rho\theta v^2v_x\,dx\end{aligned}$$

By Cauchy's inequality we have $kv_x\theta \le \frac{\lambda_1}{2}v_x^2 + \frac{k^2}{2\lambda_1}\theta^2$ so that

$$3\int_0^1 k\rho\theta v^2 v_x\,dx \le \frac{3\lambda_1}{2}\int_0^1 \rho v^2 v_x^2 dx + \frac{3k^2}{2\lambda_1}\int_0^1 \rho v^2\theta^2 dx.$$

As a result,

$$\frac{K_6}{\lambda_1}\int_0^1 v^3 v_t\,dx \le -\frac{3K_6}{2}\int_0^1 \rho v^2 v_x^2 dx + \frac{3k^2 K_6}{2\lambda_1^2}\int_0^1 \rho v^2\theta^2 dx.$$

Combining this with (6.35) yields

$$\begin{aligned} &\textstyle\int_0^1 ww_t\,dx + \frac{K_6}{\lambda_1}\int_0^1 v^3 v_t\,dx + \frac{\lambda_2}{2}\int_0^1 \rho\theta_x^2 dx \\ &\quad\le \left(K_7 + \frac{3k^2K_6}{2\lambda_1^2}\right)\textstyle\int_0^1 \rho v^2\theta^2 dx \\ &\qquad +K_4\left(1+\textstyle\int_0^1 w^2 dx\right) - \frac{K_6}{2}\textstyle\int_0^1 \rho v^2 v_x^2 dx \\ &\quad\le K_8\textstyle\int_0^1 \rho v^2\theta^2 dx + K_4\left(1+\int_0^1 w^2dx\right). \end{aligned} \tag{6.36}$$

where $K_8 = K_7 + (3k^2K_6)/(2\lambda_1^2)$.

Using Lemma 6.9 ($\int_0^1 v^2dx \le 2A_1$), Corollary 6.11 ($\rho \le A_4$), and Lemma 6.12 ($\theta^2 \le \eta J_1 + A_5 J_2 + A_6$), we have

$$\int_0^1 \rho v^2\theta^2 dx \le 2A_1A_4[\eta J_1 + A_5J_2 + A_6], \tag{6.37}$$

so (6.36) becomes

$$\begin{aligned} &\textstyle\int_0^1 ww_t\,dx + \frac{K_6}{\lambda_1}\int_0^1 v^3v_t\,dx + \frac{\lambda_2}{2}J_1 \\ &\quad \le K_9 + K_{10}\eta J_1 + K_{11}J_2 + K_4\textstyle\int_0^1 w^2dx \end{aligned}$$

where $K_9 = K_4 + 2A_1A_4A_6$, $K_{10} = 2A_1A_4$, and $K_{11} = 2A_1A_4A_5$. Since $\eta > 0$ is arbitrary, choose $\eta = \lambda_2/(4K_{10})$; then

$$\begin{aligned} &\tfrac{d}{dt}\left[\tfrac12\textstyle\int_0^1 w^2dx + \frac{K_6}{4\lambda_1}\int_0^1 v^4dx + \frac{\lambda_2}{4}J_2(t)\right] \\ &\quad\le K_9 + K_{11}J_2(t) + K_4\textstyle\int_0^1 w^2dx \\ &\quad\le K_9 + K_{12}\left[\tfrac12\textstyle\int_0^1 w^2dx + \frac{K_6}{4\lambda_1}\int_0^1 v^4dx + \frac{\lambda_2}{4}J_2(t)\right] \end{aligned} \tag{6.38}$$

where $K_{12} = \max\{4K_{11}/\lambda_2, 2K_4\}$. Define

$$Q(t) = \frac12\int_0^1 w^2dx + \frac{K_6}{4\lambda_1}\int_0^1 v^4dx + \frac{\lambda_2}{4}J_2(t);$$

then (6.38) is of the form

$$Q'(t) \le K_9 + K_{12}Q(t), \quad \text{for } t\in[0,T].$$

Thus, $[\exp(-K_{12}t)Q]' \leq K_9 \exp(-K_{12}t)$ which implies

$$Q(t) \leq e^{K_{12}t}[Q(0) + K_9 \int_0^t e^{-K_{12}s} ds\,] \leq e^{K_{12}T}[Q(0) + K_9 T] =: K_{13}.$$

For $K_{14} = \min\{1/2, K_6/(4\lambda_1), \lambda_2/4\}$, we have

$$\int_0^1 w^2 dx + \int_0^1 v^4 dx + J_2(t) \leq \frac{1}{K_{14}} Q(t) \leq \frac{K_{13}}{K_{14}} =: A_7$$

for $t \in [0,T]$. Clearly $J_2(t) \leq A_7$ □

Corollary 6.14 *There are positive constants A_7 and A_8 such that*

$$\int_0^1 \theta^2 dx \leq A_7 \quad and \quad \int_0^t \int_0^1 \rho v^2 v_x^2 dx\, d\tau \leq A_8$$

for all $t \in [0,T]$.

Proof. Note that $0 \leq \theta = w - \frac{1}{2}v^2 \leq w$, so $\theta^2 \leq w^2$ and so $\int_0^1 \theta^2 dx \leq \int_0^1 w^2 dx \leq A_7$. To prove the other bound, we had in equation (6.36)

$$\begin{aligned} Q'(t) &= \tfrac{d}{dt}\left[\tfrac{1}{2}\int_0^1 w^2 dx + \tfrac{K_6}{4\lambda_1}\int_0^1 v^4 dx + \tfrac{\lambda_2}{4} J_2(t)\right] \\ &\leq K_8 \int_0^1 \rho v^2 \theta^2 dx + K_4\left(1 + \int_0^1 w^2 dx\right) - \tfrac{K_6}{2}\int_0^1 \rho v^2 v_x^2 dx. \end{aligned}$$

Using (6.37) we have

$$\begin{aligned} \tfrac{K_6}{2}\int_0^1 \rho v^2 v_x^2 dx + Q'(t) &\leq K_8\left[2A_1 A_4 \left(\eta J_1(t) + A_5 J_2(t) + A_6\right)\right] \\ &\quad + K_4(1 + A_7) \\ &\leq \alpha J_1(t) + \beta \end{aligned}$$

where $\alpha = 2A_1A_4K_8\eta$ and $\beta = K_8A_5A_7 + 2K_8A_1A_4A_6 + K_4(1 + A_7)$. Integrate with respect to t to obtain

$$\begin{aligned} \tfrac{K_6}{2}\int_0^t\int_0^1 \rho v^2 v_x^2 dx\, d\tau &\leq \tfrac{K_6}{2}\int_0^t\int_0^1 \rho v^2 v_x^2 dx\, d\tau + Q(t) \\ &\leq Q(0) + \alpha J_2(t) + \beta t \\ &\leq Q(0) + \alpha A_7 + \beta T \\ &=: \gamma. \end{aligned}$$

Thus, $\int_0^t\int_0^1 \rho v^2 v_x^2 dx\, d\tau \leq 2\gamma/K_6 =: A_8$. □

Corollary 6.15 *There is a positive constant* A_9 *such that*

$$\rho(x,t) \geq A_9$$

for all $(x,t) \in R_T$.

Proof. From Lemma 6.12 we had $\theta^2(x,t) \leq \eta J_1(t) + A_5 J_2(t) + A_6$ for $(x,t) \in \overline{\Omega} \times [0,T]$. Thus, by Corollary 6.14,

$$\theta^2(x,t) \leq \eta J_1(t) + A_5 A_7 + A_6$$

and

$$\int_0^t \theta^2 d\tau \leq \eta J_2(t) + (A_5 A_7 + A_6)t \leq \eta A_7 + (A_5 A_7 + A_6)T.$$

Using Cauchy's inequality, we have $\theta \leq \frac{1}{2}(1+\theta^2)$, so

$$\int_0^t \theta d\tau \leq \frac{1}{2}\left(t + \int_0^t \theta^2 d\tau\right) \leq \frac{1}{2}\left[T + \eta A_7 + T(A_5 A_7 + A_6)\right] =: \alpha.$$

From equation (6.28) we have

$$\begin{aligned} m_\rho(t) &\geq \tfrac{1}{A_2 A_3}\left[m_{\rho 0}^{-1} + \tfrac{k A_2 A_3}{\lambda_1} \int_0^t \theta(x,\tau)\, d\tau\right]^{-1} \\ &\geq \tfrac{1}{A_2 A_3}\left(m_{\rho 0}^{-1} + \tfrac{k A_2 A_3}{\lambda_1}\alpha\right) \\ &=: A_9 > 0 \end{aligned}$$

for all $t \in [0,T]$. □

Corollary 6.16 *There are positive constants* A_{10} *and* A_{11} *such that*

$$\int_0^t \int_0^1 \theta_x^2 dx\, d\tau \leq A_{10} \quad and \quad \int_0^t \int_0^1 v^2 v_x^2 dx\, d\tau \leq A_{11}$$

for all $t \in [0,T]$.

Proof. From Corollary 6.14 and Corollary 6.15 we have

$$A_9 \int_0^t \int_0^1 \theta_x^2 dx\, d\tau \leq \int_0^t \int_0^1 \rho\theta_x^2 dx\, d\tau = J_2(t) \leq A_7$$

and

$$A_9 \int_0^t \int_0^1 v^2 v_x^2 dx \leq \int_0^t \int_0^1 \rho v^2 v_x^2 dx\, d\tau \leq A_8.$$

The result follows where $A_{10} = A_7/A_9$ and $A_{11} = A_8/A_9$. □

We have established the bounds: $A_9 \leq \rho(x,t) \leq A_4$, $\theta(x,t) \geq 0$, and $0 \leq z(x,t) \leq M_{z_0}$ for $(x,t) \in \overline{\Omega} \times [0,T]$. We will now establish global *a priori* bounds on the velocity function $v(x,t)$.

6.5 Velocity Bounds

Lemma 6.17 *Let $g(x,t) = \lambda_1 v_x - k\theta$; then*

$$\|\rho^{1/2} g\|^2_{L^2(\Omega)} + \int_0^t \|(\rho g)_x\|^2_{L^2(\Omega)}\, dt \le A_{12}$$

for some positive constant A_{12}.

Proof. Note that from (6.20b) we have $v_t = (\rho g)_x$. Consider

$$\begin{aligned} &\tfrac{1}{2}\tfrac{d}{dt}\textstyle\int_0^1 \rho g^2 dx \\ &\quad= \textstyle\int_0^1 \rho g g_t\, dx + \tfrac{1}{2}\int_0^1 \rho_t g^2 dx \\ &\quad= -k\textstyle\int_0^1 \rho g \theta_t\, dx + \lambda_1 \int_0^1 \rho g v_{xt}\, dx + \tfrac{1}{2}\int_0^1 \rho_t g^2 dx. \end{aligned} \tag{6.39}$$

Note that

$$\int_0^1 (\rho g) v_{xt}\, dx = (\rho g) v_t|_{x=0}^{x=1} - \int_0^1 (\rho g)_x v_t\, dx = -\int_0^1 [(\rho g)_x]^2\, dx$$

where $v_t(\partial\Omega, t) = 0$ since $v(\partial\Omega, t) = 0$. Replacing this in (6.39) yields

$$\frac{1}{2}\frac{d}{dt}\int_0^1 \rho g^2 dx + \lambda_1 \int_0^1 [(\rho g)_x]^2\, dx = -k\int_0^1 \rho g \theta_t\, dx + \frac{1}{2}\int_0^1 g^2 \rho_t\, dx.$$

Using (6.20a,c), we get

$$\begin{aligned} &\tfrac{1}{2}\tfrac{d}{dt}\textstyle\int_0^1 \rho g^2 dx + \lambda_1 \int_0^1 [(\rho g)_x]^2\, dx \\ &\quad= -(k+\tfrac{1}{2})\textstyle\int_0^1 \rho^2 g^2 v_x\, dx - k\lambda_2 \int_0^1 \rho g(\rho\theta_x)_x\, dx \\ &\qquad -k\delta \textstyle\int_0^1 \rho g f(\rho,\theta,z)\, dx \\ &\quad=: I_1 + I_2 + I_3. \end{aligned} \tag{6.40}$$

We now analyze each integral in (6.40).

Integral I_1. Integrate I_1 by parts and use the boundary conditions on v in (6.22) to obtain

$$\begin{aligned} I_1 &= -(k+\tfrac{1}{2})\textstyle\int_0^1 (\rho g)^2 v_x\, dx \\ &= -(k+\tfrac{1}{2})\left[(\rho g)^2 v|_{x=0}^{x=1} - \textstyle\int_0^1 2(\rho g)(\rho g)_x v\, dx\right] \\ &= (2k+1)\textstyle\int_0^1 (\rho g v)(\rho g)_x\, dx. \end{aligned}$$

Using Cauchy's inequality we have

$$(2k+1)(\rho g v)(\rho g)_x \le \frac{(2k+1)^2}{\lambda_1}\rho^2 g^2 v^2 + \frac{\lambda_1}{4}[(\rho g)_x]^2.$$

Also, $g^2 = (\lambda_1 v_x - k\theta)^2 \le 2(\lambda_1^2 v_x^2 + k^2\theta^2)$ since $(a-b)^2 \le 2(a^2+b^2)$. Thus,

$$(2k+1)(\rho g v)(\rho g)_x \le \frac{2A_4^2(2k+1)^2}{\lambda_1} \rho(\lambda_1^2 v_x^2 + k^2\theta^2)v^2 + \frac{\lambda_1}{4}\left[(\rho g)_x\right]^2$$

where we have used the bound $\rho \le A_4$. Integrate to obtain

$$I_1 \le K_1 \int_0^1 v^2 v_x^2 dx \, + K_2 \int_0^1 v^2\theta^2 dx \, + \frac{\lambda_1}{4} \int_0^1 \left[(\rho g)_x\right]^2 dx$$

where $K_1 = 2\lambda_1 A_4^2(2k+1)^2$ and $K_2 = 2A_4^2 k^2(2k+1)^2/\lambda_1$. Using equation (6.37) and the bound on J_2, we have

$$\int_0^1 v^2\theta^2 dx \, \le \alpha J_1(t) + \beta \le \alpha A_4 \int_0^1 \theta_x^2 dx \, + \beta$$

for constants $\alpha = 2A_1\eta$ and $\beta = 2A_1(A_5A_7 + A_6)$ so the inequality for I_1 becomes

$$I_1 \le K_1 \int_0^1 v^2 v_x^2 dx \, + K_3 \int_0^1 \theta_x^2 dx \, + K_4 + \frac{\lambda_1}{4} \int_0^1 \left[(\rho g)_x\right]^2 dx$$

where $K_3 = \alpha A_4 K_2$ and $K_4 = \beta$.

Integral I_2. Integrate I_2 by parts and use the boundary conditions on θ_x in (6.22) to obtain

$$\begin{aligned} I_2 &= -k\lambda_2 \textstyle\int_0^1 (\rho g)(\rho\theta_x)_x \, dx \\ &= -k\lambda_2 \left[(\rho g)(\rho\theta_x)|_{x=0}^{x=1} - \textstyle\int_0^1 (\rho g)_x(\rho\theta_x)\, dx \right] \\ &= k\lambda_2 \textstyle\int_0^1 (\rho g)_x(\rho\theta_x)\, dx. \end{aligned}$$

Using Cauchy's inequality we have

$$k\lambda_2(\rho g)_x(\rho\theta_x) \le \frac{k^2\lambda_2^2}{\lambda_1}(\rho\theta_x)^2 + \frac{\lambda_1}{4}\left[(\rho g)_x\right]^2.$$

Using the upper bound on ρ and integrating, we get

$$I_2 \le K_5 \int_0^1 \theta_x^2 dx \, + \frac{\lambda_1}{4} \int_0^1 \left[(\rho g)_x\right]^2 dx$$

where $K_5 = k^2\lambda_2^2 A_4^2/\lambda_1$.

Integral I_3. Since f is globally Lipschitz continuous and since ρ and z are bounded, we have

$$f(\rho,\theta,z) \le f(\rho,0,z) + K_6\theta \le K_7 + K_6\theta \le K_8(\theta+1)$$

where K_6 is the Lipschitz constant for f, K_7 is the bound on $f(\rho, 0, z)$ for $(\rho, z) \in [0, A_4] \times [0, M_{z_0}]$, and $K_8 = \max\{K_6, K_7\}$. Consequently,

$$f^2(\rho, \theta, z) \le K_8^2(\theta + 1)^2 \le 2K_8^2(\theta^2 + 1)$$

since $(a+b)^2 \le 2(a^2 + b^2)$. Using Cauchy's inequality and this last fact, we have $-k\delta g f \le \frac{1}{2}k^2\delta^2 g^2 + \frac{1}{2}f^2$ so that an integration yields

$$\begin{aligned} I_3 &\le \tfrac{1}{2}k^2\delta^2 \textstyle\int_0^1 \rho g^2 dx + \tfrac{1}{2}\int_0^1 \rho f^2 dx \\ &\le K_9 \textstyle\int_0^1 \rho g^2 dx + K_8^2 A_4 \int_0^1 (\theta^2 + 1)\, dx \\ &\le K_9 \textstyle\int_0^1 \rho g^2 dx + K_{10} \end{aligned}$$

where $K_9 = \frac{1}{2}k^2\delta^2$ and $K_{10} = K_8^2 A_4(A_1 + 1)$ (and where we have used $\int_0^1 \theta^2 dx \le A_1$ from Lemma 6.9).

Combining these results in (6.40), we obtain

$$\begin{aligned} &\tfrac{1}{2}\tfrac{d}{dt}\textstyle\int_0^1 \rho g^2 dx + \tfrac{\lambda_1}{2}\int_0^1 [(\rho g)_x]^2\, dx \\ &\le K_{11} + K_{12}\textstyle\int_0^1 \theta_x^2 dx + K_1 \int_0^1 v^2 v_x^2 dx + K_9 \int_0^1 \rho g^2 dx \end{aligned}$$

where $K_{11} = K_4 + K_{10}$ and $K_{12} = K_3 + K_5$. Integrate with respect to t to obtain

$$\begin{aligned} \textstyle\int_0^1 \rho g^2 dx &\le \textstyle\int_0^1 \rho g^2 dx + \lambda_1 \int_0^t \int_0^1 [(\rho g)_x]^2\, dx\, d\tau \\ &\le \textstyle\int_0^1 \rho_0 g_0^2 dx + 2K_{11}t + 2K_{12}\int_0^t\int_0^1 \theta_x^2 dx\, d\tau \\ &\quad + 2K_1 \textstyle\int_0^t\int_0^1 v^2 v_x^2 dx\, d\tau + 2K_9 \int_0^t \int_0^1 \rho g^2 dx\, d\tau \\ &\le \textstyle\int_0^1 \rho_0 g_0^2 dx + 2K_{11}T + 2K_{12}A_{11} + 2K_1 A_{12} \\ &\quad + 2K_9 \textstyle\int_0^t\int_0^1 \rho g^2 dx\, d\tau \\ &= K_{13} + 2K_9 \textstyle\int_0^t\int_0^1 \rho g^2 dx\, d\tau \end{aligned} \tag{6.41}$$

where we have used the bounds A_{10} and A_{11} constructed in Corollary 6.16. Equation (6.41) is a Gronwall's inequality, so we have

$$\int_0^1 \rho g^2 dx \le K_{13}\exp(2K_9 t) \le K_{13}\exp(2K_9 T) =: K_{14}.$$

Replacing this in (6.41) gives us

$$\|\rho^{1/2} g\|_{L^2(\Omega)}^2 + \lambda_1 \int_0^t \|(\rho g)_x\|_{L^2(\Omega)}^2\, dx \le A_{12}$$

where $A_{12} = K_{13} + 2K_9 K_{14} T$. □

Corollary 6.18 *There is a positive constant A_{13} such that*

$$|v(x,t| \le A_{13}$$

for all $(x,t) \in R_T$.

Proof. From Lemma 6.17 we have

$$\int_0^1 (\lambda_1 v_x - k\theta)^2 dx \le \frac{1}{A_9}\int_0^1 \rho(\lambda_1 v_x - k\theta)^2 dx \le \frac{A_{12}}{A_9}$$

where $0 < A_9 \le \rho$ (from Corollary 6.15). Using Cauchy's inequality, we have

$$\lambda_1^2 v_x^2 \le 2(\lambda_1 v_x - k\theta)^2 + 2k^2\theta^2.$$

Dividing by λ_1^2 and integrating yields

$$\begin{aligned} \textstyle\int_0^1 v_x^2 dx &\le \textstyle\frac{2}{\lambda_1^2}\int_0^1 (\lambda_1 v_x - k\theta)^2 dx + \frac{2k^2}{\lambda_1}\int_0^1 \theta^2 dx \\ &\le \textstyle\frac{2A_{12}}{\lambda_1^2 A_9} + \frac{2k^2 A_1}{\lambda_1^2} =: A_{13}^2 \end{aligned} \tag{6.42}$$

where we have used Lemma 6.9 ($\int_0^1 \theta^2 dx \le A_1$). Consequently,

$$|v(x,t)| = \left|\int_0^x v_x(\xi,t)\,d\xi\right| \le \int_0^1 |v_x|\,dx \le \left[\int_0^1 v_x^2 dx\right]^{1/2} \le A_{13}$$

where we have used Hölder's inequality. □

Corollary 6.19 *There are positive constants A_{14} and A_{15} such that*

$$\|\rho_t(\cdot,t)\|^2_{L^2(\Omega)} \le A_{14} \quad \textit{and} \quad \|\rho_x(\cdot,t)\|^2_{L^2(\Omega)} \le A_{15}$$

for $t \in [0,T]$. Thus, the function $\rho(x,t)$ is Hölder continuous on the set R_T.

Proof. In Corollary 6.18 we had the estimate (6.42): $\|v_x(\cdot,t)\|_{L^2(\Omega)} \le A_{13}$. From (6.20a), we have

$$\|\rho_t(\cdot,t)\|^2_{L^2(\Omega)} = \|\rho^2(\cdot,t)v_x(\cdot,t)\|^2_{L^2(\Omega)} \le A_4^4 A_{13}^2 =: A_{14} \tag{6.43}$$

where we have used the upper bound on ρ.

If one differentiates (6.27) with respect to x, then

$$\rho_x = \rho(v_0 - v) - \rho^2 Y^{-1} B^{-1}\left[\frac{\partial}{\partial x}(\rho_0^{-1}) + \frac{k}{\lambda_1}\int_0^t YB(\theta_x + (v_0 - v)\theta)\,d\tau\right].$$

Using the bounds on ρ, $\int_0^1 v\,dx$, $\int_0^1 \theta_x^2\,dx$, $\int_0^1 v^2\theta^2\,dx$, Y, and B, one can eventually obtain the bound

$$\|\rho_x(\cdot,t)\|^2_{L^2(\Omega)} = \int_0^1 \rho_x^2 dx \le A_{15} \tag{6.44}$$

for some constant A_{15}. It follows from (6.43) and (6.44) that ρ is Hölder continuous on R_T. □

Lemma 6.20 *There is a positive constant* A_{16} *such that*

$$\|v_x(\cdot,t)\|^2_{L^2(\Omega)} + \int_0^t \|v_{xx}(\cdot,t)\|^2_{L^2(\Omega)}dt + \int_0^t \|v_t(\cdot,t)\|^2_{L^2(\Omega)}dt \le A_{16}$$

for $t \in [0,T]$.

Proof. Since $v_t = (\rho g)_x$, Lemma 6.17 gives us $\|v_t(\cdot,t)\|^2_{L^2(\Omega)} \le A_{12}$. In Corollary 6.18, equation (6.42), we have $\|v_x(\cdot,t)\|^2_{L^2(\Omega)} \le A_{13}^2$. We only need to find a bound for $\int_0^t \|v_{xx}(\cdot,t)\|^2 dt$.

Using the identity: $a^2 \le 2(a-b)^2 + b^2$, we have $\lambda_1^2 v_{xx}^2 \le 2(\lambda_1 v_{xx} - k\theta_x)^2 + k^2\theta_x^2$. Integrating, we have

$$\int_0^t \int_0^1 v_{xx}^2 dx\, d\tau \le K_1 \int_0^t \int_0^1 g_x^2 dx\, d\tau + K_2 \int_0^t \int_0^1 \theta_x^2 dx\, d\tau \qquad (6.45)$$

where $K_1 = 2/\lambda_1^2$ and $K_2 = k^2/\lambda_1^2$. Using the identity: $a^2 \le 2(a+b)^2 + b^2$, we have

$$(\rho g_x)^2 \le 2(\rho g_x + \rho_x g)^2 + (\rho_x g)^2 = 2\left[(\rho g)_x\right]^2 + (\rho_x g)^2.$$

Integrating, we have

$$\begin{aligned} \int_0^t \int_0^1 g_x^2 dx\, d\tau &\le \tfrac{1}{A_9^2} \int_0^t \int_0^1 (\rho g_x)^2 dx\, d\tau \\ &\le \tfrac{2}{A_9^2} \int_0^t \int_0^1 \left[(\rho g)_x\right]^2 dx\, d\tau + \int_0^t \int_0^1 (\rho_x g)^2 dx\, d\tau \end{aligned}$$

where we have used the lower bound on the density function. Combining this with equation (6.45) gives us

$$\begin{aligned} \int_0^t \int_0^1 v_{xx}^2 dx\, d\tau &\le K_3 \int_0^t \int_0^1 \left[(\rho g)_x\right]^2 dx\, d\tau + K_2 \int_0^t \int_0^1 \theta_x^2 dx\, d\tau \\ &\quad + K_4 \int_0^t \int_0^1 (\rho_x g)^2 dx\, d\tau \end{aligned}$$

where $K_3 = 2K_1/A_9^2$ and $K_4 = K_1/A_9^2$. Using Lemma 6.17 and Corollary 6.16, we have

$$\int_0^t \int_0^1 v_{xx}^2 dx\, d\tau \le K_5 + K_4 \int_0^t \int_0^1 (\rho_x g)^2 dx\, d\tau \qquad (6.46)$$

where $K_5 = K_3 A_{12} + K_2 A_{10}$.

Finally,

$$\begin{aligned} \int_0^t \int_0^1 (\rho_x g)^2 dx\, d\tau &\le \max_{[0,T]} \|p_x(\cdot,t)\|^2_{L^2(\Omega)} \int_0^t \int_0^1 g^2 dx \\ &\le A_{15} \tfrac{A_{12}}{A_9} =: K_6 \end{aligned}$$

where we have used (6.44) and Lemma 6.17. Combining this with (6.46) gives the bound

$$\int_0^t \|v_{xx}(\cdot,t)\|^2_{L^2(\Omega)}\,d\tau \le A_{16}$$

where $A_{16} = K_5 + K_4K_6$. □

6.6 Temperature Bounds

We have constructed bounds on the density $\rho(x,t)$, the velocity $v(x,t)$, and the fuel concentration $z(x,t)$. We now finish with the *a priori* bounds on the temperature $\theta(x,t)$. The upper bound on temperature is derived from the lower bound on density and a comparison theorem.

Lemma 6.21 *There is a positive constant A_{17} such that*

$$\theta(x,t) \ge A_{17}$$

for all $(x,t) \in R_T$.

Proof. From (6.20c) we have

$$\begin{aligned}
\theta_t &= \lambda_2(\rho\theta_x)_x + \lambda_1\rho v_x^2 - k\rho\theta v_x + \delta f(\rho,\theta,z)\\
&\ge \lambda_2(\rho\theta_x)_x + \rho(\lambda_1 v_x^2 - k\theta v_x)\\
&= \lambda_2(\rho\theta_x)_x + \rho\left[\lambda_1(v_x - \tfrac{k\theta}{2\lambda_1})^2 - \tfrac{k^2\theta^2}{4\lambda_1}\right]\\
&\ge \lambda_2(\rho\theta_x)_x - \tfrac{\rho k^2}{4\lambda_1}\theta^2\\
&\ge \lambda_2(\rho\theta_x)_x - \tfrac{k^2A_4}{4\lambda_1}\theta^2 \quad \text{by Corollary 6.11.}
\end{aligned}$$

Let $\phi(t)$ be the solution to:

$$\frac{d\phi}{dt} = -\frac{k^2A_4}{4\lambda_1}\phi^2,\ t>0,\ \ \phi(0) = m_{\theta_0};$$

then $\phi(t) = (4\lambda_1 m_{\theta_0})/(4\lambda_1 + k^2A_4t)$,

$$\theta_t - \lambda_2(\rho\theta_x)_x + \frac{k^2A_4}{4\lambda_1}\theta^2 \ge \phi_t - \lambda_2(\rho\phi_x)_x + \frac{k^2A_4}{4\lambda_1}\phi^2 \equiv 0$$

for $(x,t) \in \Omega\times(0,T)$, and on the parabolic boundary, $\theta(x,0) = \theta_0(x) \ge m_{\theta_0} = \phi(0)$ for $x \in \overline{\Omega}$ and $\theta_x(\partial\Omega,t) = 0 = \phi_x(0)$ for $t \ge 0$. By the maximum principle, $\theta(x,t) \ge \phi(t)$ for all $(x,t) \in \overline{\Omega}\times[0,T)$. Since $\phi(t) \ge \phi(T) =: A_{17}$, we have $\theta(x,t) \ge A_{17} > 0$ for $(x,t) \in R_T$. □

Lemma 6.22 *There is a positive constant A_{18} such that*

$$\|\theta_x(\cdot,t)\|^2_{L^2(\Omega)} + \int_0^t \|\theta_{xx}(\cdot,t)\|^2_{L^2(\Omega)}\,dt + \int_0^t \|\theta_t(\cdot,t)\|^2_{L^2(\Omega)}\,dt \le A_{18}$$

for $t \in [0,T]$.

Proof. Using (6.20c) and the boundary conditions (6.22), we have

$$\begin{aligned} \tfrac{1}{2}\tfrac{d}{dt}\textstyle\int_0^1 \theta_x^2\,dx &= \textstyle\int_0^1 \theta_x\theta_{xt}\,dx \\ &= \theta_x\theta_t|_{x=0}^{x=1} - \textstyle\int_0^1 \theta_t\theta_{xx}\,dx \\ &= -\textstyle\int_0^1 \theta_{xx}\left[\lambda_2(\rho\theta_x)_x + \rho(\lambda_1 v_x - k\theta)v_x + \delta f(\rho,\theta,z)\right]dx \\ &= -\lambda_2 \textstyle\int_0^1 \rho\theta_{xx}^2\,dx - \lambda_2\int_0^1 \rho_x\theta_x\theta_{xx}\,dx \\ &\quad - \textstyle\int_0^1 \rho(\lambda_1 v_x - k\theta)v_x\theta_{xx}\,dx - \delta\int_0^1 \theta_{xx} f(\rho,\theta,z)\,dx. \end{aligned} \tag{6.47}$$

Using Hölder's inequality and Young's inequality, we now observe that

$$\theta_x(x,t) = \int_0^x 2\theta_x\theta_{xx}\,dx \le 2\left(\int_0^1 \theta_x^2\,dx\right)^{1/2}\left(\int_0^1 \theta_{xx}^2\,dx\right)^{1/2}$$

so that

$$\max_{\overline{\Omega}}|\theta_x(x,t)| \le \sqrt{2}\left(\int_0^1 \theta_x^2\,dx\right)^{1/4}\left(\int_0^1 \theta_{xx}^2\,dx\right)^{1/4}$$

and

$$\begin{aligned} &-\lambda_2 \textstyle\int_0^1 \rho_x\theta_x\theta_{xx}\,dx \\ &\le \lambda_2 \left|\textstyle\int_0^1 \rho_x\theta_x\theta_{xx}\,dx\right| \\ &\le \lambda_2 \max_{\overline{\Omega}}|\theta_x| \left(\textstyle\int_0^1 \frac{\rho_x^2}{\rho}\,dx\right)^{1/2}\left(\int_0^1 \rho\theta_{xx}^2\,dx\right)^{1/2} \\ &\le \lambda_2\sqrt{2}\,A_9^{-1/4}\left(\textstyle\int_0^1 \frac{\rho_x^2}{\rho}\,dx\right)^{1/2}\left(\int_0^1 \theta_x^2\,dx\right)^{1/4}\left(\int_0^1 \rho\theta_{xx}^2\,dx\right)^{3/4} \\ &\le \lambda_2 K_1\left(\textstyle\int_0^1 \theta_x^2\,dx\right)^{1/4}\left(\int_0^1 \rho\theta_{xx}^2\,dx\right)^{3/4} \\ &\le \tfrac{\lambda_2}{8}\textstyle\int_0^1 \rho\theta_{xx}^2\,dx + K_2\int_0^1 \theta_x^2\,dx \end{aligned} \tag{6.48}$$

where $K_1 = \sqrt{2} A_{15} A_9^{3/4}$ and $K_2 = 54 K_1^4 \lambda_2$. Note that we have used (6.44) and the bounds on ρ. We also have

$$\begin{aligned}
&- \int_0^1 \rho(\lambda_1 v_x - k\theta) v_x \theta_{xx}\, dx \\
\leq\ & \left| \int_0^1 \rho(\lambda_1 v_x - k\theta) v_x \theta_{xx}\, dx \right| \\
\leq\ & A_9^{-\frac{1}{2}} \max_{\overline{\Omega}} |v_x(x,t)| \times \\
& \times \left(\int_0^1 \rho(\lambda_1 v_x - k\theta)^2\, dx \right)^{\frac{1}{2}} \left(\int_0^1 \rho \theta_{xx}^2\, dx \right)^{\frac{1}{2}} \\
\leq\ & \sqrt{\tfrac{2}{A_9}} \left(\int_0^1 v_x^2 dx \right)^{\frac{1}{4}} \times \\
& \times \left(\int_0^1 v_{xx}^2 dx \right)^{\frac{1}{4}} \left(\int_0^1 \rho(\lambda_1 v_x - k\theta)^2 dx \right)^{\frac{1}{2}} \left(\int_0^1 \rho \theta_{xx}^2 dx \right)^{\frac{1}{2}} \\
\leq\ & K_3 \left(\int_0^1 v_{xx}^2\, dx \right)^{\frac{1}{4}} \left(\int_0^1 \rho \theta_{xx}^2\, dx \right)^{\frac{1}{2}} \\
& \leq K_4 \left(\int_0^1 v_{xx}^2\, dx \right)^{\frac{1}{2}} + \tfrac{\lambda_2}{8} \int_0^1 \rho \theta_{xx}^2\, dx
\end{aligned} \tag{6.49}$$

where $K_3 = (2A_{12}/A_9)^{1/2} A_{16}^{1/4}$ and $K_4 = 2K_3^2/\lambda_2$. Note that we have used Lemma 6.17 and Lemma 6.20. Finally,

$$\begin{aligned}
&-\delta \int_0^1 \theta_{xx} f(\rho, \theta, z)\, dx \\
&\leq K_5 \int_0^1 |\theta_{xx}(1+\theta)|\, dx \\
&\leq K_5 \left(\int_0^1 \rho \theta_{xx}^2\, dx \right)^{1/2} \left(\int_0^1 \tfrac{(1+\theta)^2}{\rho} \right)^{1/2} \\
&\leq \tfrac{\lambda_2}{4} \int_0^1 \rho \theta_{xx}^2\, dx + \tfrac{1}{A_9} \lambda_2 \int_0^1 (1+\theta)^2 dx \\
&\leq \tfrac{\lambda_2}{4} \int_0^1 \rho \theta_{xx}^2\, dx + K_6
\end{aligned} \tag{6.50}$$

where $f(\rho, \theta, z) \leq f(\rho, 0, z) + L\theta \leq K_5(1+\theta)$ and $K_6 = 2(1 + 2A_1 + A_7)/(A_9 \lambda_2)$. Note that we have used the bounds on $\int_0^1 \theta\, dx$ and $\int_0^1 \theta^2\, dx$.

Replacing equations (6.48) through (6.49) into (6.50) gives us

$$\begin{aligned}
&\tfrac{1}{2} \tfrac{d}{dt} \int_0^1 \theta_x^2\, dx + \tfrac{\lambda_2}{2} \int_0^1 \rho \theta_{xx}^2\, dx \\
&\leq K_2 \int_0^1 \theta_x^2\, dx + K_4 \left(\int_0^1 v_{xx}^2\, dx \right)^{1/2} + K_6.
\end{aligned}$$

Integrating with respect to t gives us

$$\begin{aligned}
&\tfrac{1}{2} \int_0^1 \theta_x^2\, dx + \tfrac{\lambda_2}{2} \int_0^t \int_0^1 \rho \theta_{xx}^2\, dx \\
&\leq \tfrac{1}{2} \int_0^1 \theta_{0,x}^2\, dx + K_2 \int_0^t \int_0^1 \theta_x^2\, dx\, d\tau \\
&+ K_4 \int_0^t \|v_{xx}(\cdot, t)\|_{L^2}\, d\tau + K_6 t
\end{aligned}$$

from which the result

$$\int_0^1 \theta_x^2\, dx + \int_0^t \int_0^1 \theta_{xx}^2\, dx\, d\tau \leq K_7$$

immediately follows. From the other previously derived bounds, we have

$$\int_0^t \int_0^1 \theta_t^2 \, d\tau \leq K_8$$

and the lemma is proved (where $A_{18} = K_7 + K_8$). □

The upper bound on temperature is immediate from the bounds obtained on the L^2 norm of θ_x (*c.f.* Corollary 6.18).

Corollary 6.23 *There is a positive constant* A_{19} *such that*

$$\theta(x,t) \leq A_{19}$$

for all $(x,t) \in R_T$.

These *a priori* bounds establish the proof of Theorem 6.4. To prove Theorem 6.5, we need one final estimate involving the concentration $z(x,t)$.

Lemma 6.24 *There is a positive constant* A_{20} *such that*

$$\|z_x(\cdot,t)\|_{L^2(\Omega)}^2 + \int_0^t \|z_{xx}(\cdot,t)\|_{L^2(\Omega)}^2 \, d\tau + \int_0^t \|z_t(\cdot,t)\|_{L^2(\Omega)}^2 \, d\tau \leq A_{20}$$

for $t \in [0,T]$.

Proof. From (6.20d), it follows that

$$\begin{aligned} &\tfrac{1}{2}\tfrac{d}{dt}\textstyle\int_0^1 z_x^2 dx + \lambda_3 \int_0^1 \rho^2 z_{xx}^2 dx \\ &\quad = -2\lambda_3 \textstyle\int_0^1 \rho\rho_x z_x z_{xx} dx + \int_0^1 z_{xx} f(\rho,\theta,z)\, dx. \end{aligned}$$

Since ρ, θ, and z are bounded, we have $f(\rho,\theta,z)$ bounded. The term $\int_0^1 \rho_x z_x z_{xx}\, dx$ can be bounded in exactly the same way as $\int_0^1 \rho_x \theta_x \theta_{xx}\, dx$ in Lemma 6.22. □

6.7 Comments

Existence for gas dynamic systems (6.1) with various initial and boundary conditions is surveyed in Kazhikov and Solonnikov [KZH2] and Matsumara and Nishida [MAT]. Local existence is reasonably well understood, but global existence in higher spatial dimensions remains an important unsolved problem.

Invariance techniques as discussed in Chapter 4 fail in most cases because invariant regions, if they exist, are unbounded. The *a priori* boundedness of solutions is obtained instead by use of energy estimates. This method is illustrated for the nonreactive model (6.2)-(6.3) in Section 6.1. The global existence theorem proved is due to Kanal [KAN].

The induction model (6.7)-(6.8) in Section 6.2 with viscosity matrix $B = 0$ has only solutions which blow up in finite time because of the reaction term present. Our discussion is incomplete as we do not address the question of where blowup will occur. Jackson, Kapila, and Stewart [JAC1],[JAC2] have given a formal asymptotic discussion of this model. Majda and Rosales [MAJ1] consider a related problem. The generation of these hot spots as detected in this ignition model is believed to be crucial in the understanding of the *deflagration-to-detonation transition phenomena.*

In Sections 6.3 through 6.6, initial-boundary value problems corresponding to the behavior of a confined, heat-conductive, viscous, and chemically reactive gas are considered in one spatial dimension. Using estimates on the total free energy of the system, *a priori* bounds are found for the solutions which gives global existence. From a physical point of view, this shows that the heat conductivity and viscosity of the gas prevent shocks from developing for arbitrarily large Lipschitz continuous initial data. These sections are from Bebernes and Bressan [BEB6] and Bressan [BRE].

References

[ADA] R.A. Adams, *Sobolev Spaces*, Academic Press, New York, 1975.

[AMA1] H. Amann, *Fixed point equations and nonlinear eigenvalue problems in ordered Banach spaces*, SIAM Rev. 18 (1976), 620-709.

[AMA2] H. Amann, *Supersolutions, monotone iteration, and stability*, J. Diff. Eq., 21 (1976), 363-377.

[AMA3] H. Amann, *Invariant sets and existence theorems for semilinear parabolic and elliptic systems*, J. Math. Anal. Appl. 65 (1978), 432-467.

[AMA4] H. Amann, *Existence and stability of solutions for semilinear parabolic systems and applications to some diffusion-reaction equations*, Proc. Royal Soc. Edin. 81A (1978), 35-47.

[BAL] J. Ball, *Remarks on blowup and nonexistence theorems for nonlinear evolution equations*, Quart. J. Math. Oxford 28 (1977), 473-486.

[BAN1] C. Bandle, *Existence theorems, qualitative results, and* a priori *bounds for a class of nonlinear Dirichlet problems*, Arch. Rat. Mech. Anal. 58 (1975), 219-238.

[BAN2] C. Bandle, *Isoperimetric Inequalities and their Applications*, Pitman, London, 1980.

[BAR1] P. Baras and L. Cohen, *Sur l'explosion totale après $T_{\max}$ de la solution d'une equation de la chaleur semi-linéaire*,C.R. Acad. Sci. Paris, t. 300 (1985), 295-298.

[BAR2] P. Baras and L. Cohen, *Complete blowup after $T_{\max}$ for the solution of a semilinear heat equation*, J. Functional Analysis 71 (1987), 142-174.

[BRN] G.I. Barenblatt, *Similarity, Self-Similarity, and Intermediate Asymptotics*, Consultants Bureau, New York, 1979.

[BEB1] J. Bebernes and K. Schmitt, *Invariant sets and the Hukuhara-Kneser property for systems of parabolic partial differential equations*, Rocky Mtn. J. Math. 7 (1977), 557-567.

[BEB2] J. Bebernes and K. Schmitt, *On the existence of maximal and minimal solutions for parabolic partial differential equations*, Proc. Amer. Math. Soc. 73 (1979), 211-218.

[BEB3] J. Bebernes, K.N. Chueh, and W. Fulks, *Some applications of invariance for parabolic systems*, Indiana Univ. Math. J. 28 (1979), 269-277.

[BEB4] J. Bebernes and D. Kassoy, *A mathematical analysis of blow-up for thermal runaway*, SIAM J. Appl. Math. 40 (1981), 476-484.

[BEB5] J. Bebernes and A. Bressan, *Thermal behavior for a confined reactive gas*, J. Diff. Equations 44 (1982), 118-133.

[BEB6] J. Bebernes and A. Bressan, *Global* a priori *estimates for a viscous reactive gas*, Proc. Royal Soc. Edinburgh 101A (1985), 321-333.

[BEB7] J. Bebernes, *Solid fuel combustion–Some mathmematical problems*, Rocky Mountain J. Math. 16 (1986), 417-433.

[BEB8] J. Bebernes, D. Eberly, and W. Fulks, *Solution profiles for some simple combustion models*, Nonlinear Anal.-Theory, Methods and Applications 10 (1986), 165-177.

[BEB9] J. Bebernes and W. Troy, *On the existence of solutions to the Kassoy problem in dimension 1*, SIAM J. Math. Anal. 18 (1987), 1157-1162.

[BEB10] J. Bebernes, A. Bressan, and D. Eberly, *A description of blow-up for the solid fuel ignition model*, Indiana Math. J. 36 (1987), 295-305.

[BEB11] J. Bebernes and D. Eberly, *A description of self-similar blowup for the solid fuel ignition model for dimensions* $n \geq 3$, Analyse Non Linéaire, Ann. Inst. Henri Poincare 5, No.1, (1988), 1-21.

[BEB12] J. Bebernes, A. Bressan, and A. Lacey, *Total blowup versus single point blowup*, J. Diff. Equations 73, No.1, (1988), 30-44.

[BEB13] J. Bebernes, A. Bressan, D.R. Kassoy, and N. Riley, *The confined nondiffusive thermal explosion with spatially homogeneous pressure variation*, to appear in Comb. Sci. Tech.

[BEL] H. Bellout, *A criterion for blow-up of solutions to semilinear heat equations*, SIAM J. Math. Anal. 18 (1987), 722-727.

[BER] M. Berger and R. Kohn, *A rescaling algorithm for the numerical calculation of blowing up solutions*, preprint.

[BRA] G. Bratu, *Sur les equations integrales non lineares*, Bull. Soc. Math. de France 42 (1914), 113-142.

[BRE] A. Bressan, *Global solutions for the one-dimensional equations of a viscous reactive gas*, Boll. U.M.I. 58 (1986), 291-308.

[BUC1] J. Buckmaster and G. Ludford, *Theory of Laminar Flames*, University Press, Cambridge, 1982.

[BUC2] J. Buckmaster and G. Ludford, *Lectures on Mathematical Combustion*, CBMS-NSF Regional Conference Series, SIAM, 1983.

[BUR] J.G. Burnell, A.A. Lacey, and G.C. Wake, *Steady states of reaction- diffusion equations, Part I: Questions of existence and continuity of solution branches*, J. Aust. Math. Soc. Ser. B. 24 (1983), 374-391.

[CAZ] T. Cazenave and P.-L. Lions, *Solutions globales d'equations de la chaleur semilineaires*, Comm. Partial Differential Equations, 9 (1984), 935- 978.

[CHA] J. Chandra and P.M. Davis, *Comparison theorems for systems of reaction- diffusion equations*, Proc. Int. Conf. "Recent Trends in Differential Equations", Trieste, Academic Press.

[CHU] K.N. Chueh, C.C. Conley, and J.A. Smoller, *Positively invariant regions for systems of nonlinear diffusion equations*, Indiana Univ. Math. J. 26 (1977), 373-392.

[CLA] J.F. Clarke and D.R. Kassoy, J. Fluid Mech. 150 (1985), 253.

[DAN1] E.N. Dancer, *On the structure of solutions of an equation in catalysis theory when a parameter is large*, J. Diff. Equations, 37 (1980), 404-437.

[DAN2] E.N. Dancer, *Uniqueness for elliptic equations when a parameter is large*, Nonlinear Analysis 8 (1984), 835-836.

[DAN3] E.N. Dancer, *On the number of positive solutions of weakly nonlinear elliptic equations when a parameter is large*, Nonlinear Analysis, to appear.

[DEF] D. de Figueiredo and P.-L.Lions, *On pairs of positive solutions for a class of semilinear elliptic problems*, Indiana Univ. J. Math. 34 (1985), 581-606.

[DIA] J.I. Diaz, *Nonlinear Partial Differential Equations and Free Boundaries*, Pitman Research Notes in Mathematics 106, London, 1985.

[DOL] J. Dold, *Analysis of the early stage of thermal runaway*, Quart. J. Mech. Appl. Math. 38 (1985), 361-387.

[EAT1] B. Eaton and K. Gustafson, *Exact solutions and ignition parameters in the Arrhenius conduction theory of gaseous thermal explosion*, J. Appl. Math. and Phys. (ZAMP) 33 (1982), 392-404.

[EAT2] B. Eaton and K. Gustafson, *Calculation of critical branching points in two-parameter bifurcation problems*, J. Comp. Phys. 50 (1983), 171-177.

[EBE1] D. Eberly and W. Troy, *On the existence of logarithmic-type solutions to the Kassoy-Kapila problem in dimensions* $3 \leq n \leq 9$, J. Diff. Eq. 70 (1987), 309-324.

[EBE2] D. Eberly, *On the nonexistence of solutions to the Kassoy problem in dimensions 1 and 2*, J. Math. Anal. Appl. 129 (1988), 401-408.

[ESC] M. Escobedo and O. Kavian, *Asymptotic behavior of positive solutions of a nonlinear heat equation*, Houston Math. J., to appear.

[FIF] P. Fife, *Mathematical Aspects of Reacting and Diffusing Systems*, Vol. 28 Lecture Notes in Biomathematics, Springer-Verlag, New York, 1979.

[FRA] D.A. Frank-Kamanetski, *Diffusion and Heat Exchange in Chemical Kinetics*, Princeton University Press, Princeton, 1955.

[FRI1] A. Friedman, *Partial Differential Equations of Parabolic Type*, Prentice-Hall, New Jersey, 1964.

[FRI2] A. Friedman and B. McLeod, *Blowup of positive solutions of semilinear heat equations*, Indiana Univ. J. Math. 34 (1985), 425-477.

[FRI3] A. Friedman and Y. Giga, *A single point blowup for solutions of semilinear parabolic systems*, J. Fac. Sci. Univ. Tokyo 34, Sect. 1A, (1987), to appear.

[FRI4] A. Friedman, J. Friedman, and B. McLeod, *Concavity of solutions of nonlinear ordinary differential equations*, preprint.

[FOI] C. Foias, O.P. Manley, and R. Temam, *New representation of Navier-Stokes equations governing self-similar homogeneous turbulence*, Phys. Rev. Lett. 51 (1983), 617-620.

[FUJ1] H. Fujita, *On the blowing up of solutions to the Cauchy problem for* $u_t = \Delta u + u^{1+\alpha}$, J. Fac. Sci. Univ. Tokyo 13, Sect. 1A, (1966), 109-124.

[FUJ2] H. Fujita and S. Watanabe, *On the uniqueness and nonuniqueness of solutions of initial value problems for some quasilinear parabolic equations*, Comm. Pure Appl. Math. 21 (1968), 563-652.

[FUJ3] H. Fujita, *On the nonlinear equations* $\Delta u + e^u = 0$ *and* $v_t = \Delta v + e^v$, Bull. Amer. Math. Soc. 75 (1969), 132-135.

[FUJ4] H. Fujita, *On some nonexistence and nonuniqueness theorems for nonlinear parabolic equations*, Proc. Symp. Pure Math. XVIII, Nonlinear Functional Analysis, Amer. Math. Soc. 28 (1970), 105-113.

[GAL1] V.A. Galaktionov and Samarskii, *Methods of constructing approximate self- similar solutions of nonlinear heat equations I-IV*, Math. USSR Sbornik 46 (1983), 291-321 (I); 46 (1983), 439-458 (II); 48 (1984), 1-18 (III); 49 (1984), 125-149 (IV).

[GAL2] V.A. Galaktionov and S.A. Posashkov, *The equation* $u_t = u_{xx} + u^\beta$. *Localization and asymptotic behavior of unbounded solutions*, Diff. Uravnen 22 (1986), 1165-1173.

[GAR] R. Gardner, *Solutions of a nonlocal conservation law arising in combustion theory*, SIAM J. Math. Anal. 18 (1987), 172-183.

[GEL] I.M. Gelfand, *Some problems in the theory of quasilinear equations*, Amer. Math. Soc. Trans. 29 (1963), 295-381.

[GID1] B. Gidas, W. Ni, and L. Nirenberg, *Symmetry and related problems via the maximum principle*, Comm. Math. Phys. 68 (1979), 209-243.

[GID2] B. Gidas and J. Spruck, A priori *bounds for positive solutions of nonlinear elliptic equations*, Comm. Partial Diff. Eq. 6 (1981), 883-901.

[GID3] B. Gidas and J. Spruck, *Global and local behavior of positive solutions of nonlinear elliptic equations*, Comm. Pure Appl. Math. 34 (1981), 525-598.

[GIG1] Y. Giga, *A bound for global solutions of semilinear heat equations*, Comm. Math. Phys. 103 (1986), 415-421.

[GIG2] Y. Giga, *Self-similar solutions for semilinear parabolic equations*, in Nonlinear Systems of Partial Differential Equations in Applied Mathematics, B. Nicolaenko *et.al.*, eds., Amer. Math. Soc. Lecture Notes Appl. Math. 23 (1986), Part 2, 247-252.

[GIG3] Y. Giga, *Solutions for semilinear parabolic equations in L^p and regularity of weak solutions of the Navier-Stokes system*, J. Diff. Equations 62 (1986), 186-212.

[GIG4] Y. Giga, *On elliptic equations related to self-similar solutions for nonlinear heat equations*, Hiroshima Math. J. 16 (1986), 541-554.

[GIG5] Y. Giga and R. Kohn, *Asymptotically self-similar blowup of semilinear heat equations*, Comm. Pure Appl. Math. 38 (1985), 297-319.

[GIG6] Y. Giga and R. Kohn, *Characterizing blowup using similarity variables*, Indiana Univ. Math. J. 36 (1987), 1-40.

[GIG7] Y. Giga and R. Kohn, *Removability of blowup points for semilinear heat equations*, preprint.

[GIL] D. Gilbarg and N. Trudinger, *Elliptic Partial Differential Equations of Second Order*, Springer-Verlag, New York, 1977.

[HAR] A. Haraux and F.B. Weissler, *Nonuniqueness for a semilinear initial value problem*, Indiana Univ. Math. J. 31 (1982), 167-189.

[HEN] D. Henry, *Geometric Theory of Semilinear Parabolic Equations*, Lecture Notes in Mathematics 840, Springer-Verlag, New York, 1981.

[HOC] L.M. Hocking, K. Stewartson, and J. Stuart, *A nonlinear instability burst in plane parallel flow*, J. Fluid Mech. 51 (1972), 702-735.

[HOL] S.L. Hollis, R.H. Martin, and M. Pierre, *Global existence and boundedness in reaction-diffusion systems*,

[JAC1] T.L. Jackson and A.K. Kapila, *Shock-induced thermal runaway*, SIAM J. Appl. Math. 45 (1985), 130-137.

[JAC2] T.L. Jackson, A.K. Kapila, and D.S. Stewart, *Evolution of a reaction center in an explosive material*, Univ. of Ill. T.& A.M. Report No. 484, Feb. 1987.

[JOS] D. Joseph and T. Lundgren, *Quasilinear Dirichlet problems driven by positive sources*, Arch. Rat. Mech. Anal. 49 (1973), 241-269.

[KAH] C.S. Kahane, *On a system of nonlinear parabolic equations arising in chemical engineering*, J. Math. Anal. Appl. 53 (1976), 343-358.

[KAM] S. Kamin and L.A. Peletier, *Large-time behavior of solutions of the heat equation with absorption*, Ann. Sc. Norm. Pisa Cl. Sci. (4) 12 (1984), 393- 408.

[KAN] Y. Kanal, *On some systems of quasilinear parabolic equations*, USSR Comp. Math. and Math. Phys. 6 (1966), 74-88.

[KAP1] A.K. Kapila, *Reactive-diffusive system with Arrhenius kinetics: Dynamics of ignition*, SIAM J. Appl. Math. 39 (1980), 21-36.

[KAP2] A.K. Kapila, D.R. Kassoy, and D.S. Stewart, *A unified formulation for diffusive and nondiffusive thermal explosion theory*, to appear in Comb. Sci. Tech.

[KPL] S. Kaplan, *On the growth of solutions of quasilinear parabolic equations*, Comm. Pure Appl. Math. 16 (1963), 327-330.

[KAS1] D. Kassoy, *The supercritical spatially homogeneous thermal explosion: Initiation to completion*, Quart. J. Mech. Appl. Math. 30 (1977), 71-89.

[KAS2] D. Kassoy and A. Liñan, *The influence of reactant consumption on the critical conditions for homogeneous thermal explosions*, Quart. J. Appl. Math. 31 (1978), 99-112.

[KAS3] D. Kassoy and J. Poland, *The thermal explosion confined by a constant temperature boundary: I. The induction period solution*, SIAM J. Appl. Math. 39 (1980), 412-430.

[KAS4] D. Kassoy and J. Poland, *The thermal explosion confined by a constant temperature boundary: II. The extremely rapid transient*, SIAM J. Appl. Math. 41 (1981), 231-246.

[KAS5] D. Kassoy and J. Poland, *The induction period of a thermal explosion in a gas between infinite parallel plates*, Combustion and Flame 50 (1983), 259-274.

[KAW1] B. Kawohl, *Global existence of large solutions to initial-boundary value problems for a viscous heat conducting one-dimensional real gas*, J. Diff. Eq. 58 (1985), 76-103.

[KAW2] B. Kawohl, *Qualitative properties of solutions to semilinear heat equations*, Expo. Math. 4 (1986), 257-270.

[KAZ] J. Kazdan and F. Warner, *Remarks on some quasilinear elliptic equations*, Comm. Pure Appl. Math 43 (1983), 1350-1366.

[KZH1] A.V. Kazhikov and V. Shelukin, *Unique global solution in time of initial- boundary value problems for one-dimensional equations of a viscous gas*, Prikl. Mat. Mech. J. Appl. Math. Mech. 41 (1977), 273-282.

[KZH2] A.V. Kazhikov and V.A. Solonnikov, *Existence theorems for the equations of motion of a compressible viscous fluid*, Ann. Rev. Fluid Mech. 13 (1981), 79-95.

[LAC1] A. Lacey, *Mathematical analysis of thermal runaway for spatially inhomogeneous reactions*, SIAM J. Appl. Math. 43 (1983), 1350-1366.

[LAC2] A. Lacey and D. Tzanetis, *Global existence and convergence to a singular steady state for a semilinear heat equation*, Proc. Royal Soc. Edinburgh 105A (1987), 289-305.

[LAC3] A. Lacey and D. Tzanetis, *Complete blowup for a semilinear heat equation with a sufficiently large initial condition*, preprint.

[LAD] O.A. Ladyzenskaja, V.A. Solonnikov, and N.N. Ural'ceva, *Linear and quasilinear equations of parabolic type*, Amer. Math. Soc., Providence, 1968.

[LER] J. Leray, *Sur le mouvement d'un liquide visqueux emplissant l'espace*, Acta Math. 63 (1934), 193-248.

[LEV1] H.A. Levine, *Some nonexistence and instability theorems for solutions of formally parabolic equations of the form* $Pu_t = -Au + F(u)$, Arch. Rat. Mech. Anal. 51 (1973), 371-386.

[LEV2] H.A. Levine, *The long time behavior of reaction-diffusion equations in unbounded domains: A survey*, Proc. 10th Dundee Conference on Ordinary and Partial Differential Equations, 1988.

[LNA1] A. Liñan and F.A. Williams, *Theory of ignition of a reactive solid by constant energy flux*, Comb. Sci. Tech. 3 (1971), 91.

[LNA2] A. Liñan and F.A. Williams, *Ignition of a reactive solid exposed to a step in surface temperature*, SIAM J. Appl. Math. 36 (1979), 587.

[LIO] J.-L. Lions and E. Magenes, *Nonhomogeneous boundary value problems and applications*, Vol. II., Springer-Verlag, New York, 1972.

[LIN] P.-L. Lions, *Positive solutions of semilinear elliptic equations*, SIAM Review 24 (1982), 441-467.

[LIU] J. Liouville, *Sur l'equation aux dérivées partielles* $\frac{\partial^2 \ln \lambda}{\partial u \partial v} \pm 2\lambda a^2 = 0$, J. de Math. 18 (1853), 71-72.

[MAJ1] A. Majda and R. Rosales, *A theory for spontaneous Mach-stem formation in reacting shock fronts, I*, SIAM J. Appl. Math. 43 (1983), 1310-1334.

[MAJ2] A. Majda and R. Rosales, *A theory for spontaneous Mach-stem formation in reacting shock fronts, II*, SIAM J. Appl. Math. 71 (1984), 117-148.

[MAJ3] A. Majda, *Compressible fluid flow and systems of conservation laws in several space variables*, Lecture Notes in Applied Mathematics 53, Springer- Verlag, New York, 1984.

[MAJ4] A. Majda and R. Rosales, *Resonantly interacting weakly nonlinear hyperbolic waves, I*, Stud. Appl. Math. 71 (1984), 149-179.

[MAJ5] A. Majda and R. Rosales, *Nonlinear mean field high frequency wave interactions in the induction zone*, SIAM J. Appl. Math. 47 (1987), 1017- 1039.

[MAJ6] A. Majda and R. Rosales, *Wave interactions in the induction zone*, SIAM J. Appl. Math. 47 (1987), 1017-1039.

[MAR] R.H. Martin, *Nonlinear operators and differential equations in Banach spaces*, Wiley, New York, 1976.

[MAT] A. Matsumura and T. Nishida, *Initial-coundary value problems for the equations of motion of compressible viscous fluids*, MSRI-Berkeley Report 008-83.

[MCL] D. McLaughlin, G. Papanicolau, C. Sulem, and P.-L. Sulem, *The focusing singularity of the cubic Schrödinger equation*, Phys. Rev. A. 34 (1986), 1200-1210.

[MEY] J.W. Meyer and A.K. Oppenheim, *Dynamic response of a plane-symmetrical exothermic reaction center*, AIAA J. 10 (1972), 1509-1513.

[MLA1] W. Mlak, *Differential inequalities of parabolic type*, Ann. Polon. Math. 3 (1957), 349-354.

[MLA2] W. Mlak, *Parabolic differential inequalities and the Chaplighin's method*, Ann. Polon. Math. 8 (1960), 139-152.

[MLA3] W. Mlak, *An example of the equation $u_t = u_{xx} + f(x,t,u)$ with distinct maximum and minimum solutions of a mixed problem*, Ann. Polon. Math. 13 (1963), 101-103.

[MUE] C.E. Mueller and F.B. Weissler, *Single-point blowup for a general semilinear heat equation*, Indiana Univ. Math. J. 34 (1985), 881-913.

[NAG1] M. Nagumo, *Über die Differentialgleichung $y'' = f(x,y,y')$*, Proc. Phys.-Math. Soc. Japan 19 (1937), 861-866.

[NAG2] M. Nagumo, A note in "Kansū-Hōteisiki", No. 15 (1939).

[NAG3] M. Nagumo, *Über das Rantwertproblem der nichtlinearen gewöhnlichen Differentialgleichung zweiter Ordnung*, Proc. Phys.-Math. Soc. Japan 24 (1942), 845-851.

[NAG4] M. Nagumo and S. Simoda, *Note sur l'inégalité différentielle concernant les équations du type parabolique*, Proc. Japan. Acad. 27 (1951), 536-639.

[NAG5] M. Nagumo, *On principally linear elliptic differential equations of second order*, Osaka Math. J. 6 (1954), 207-229.

[NAS] J. Nash, *Le probléme de Cauchy pour les équations différentielles d'un fluide générals*, Bull. Soc. Math. France 90 (1962), 487-497.

[NI] W.-M. Ni, P.E. Sacks, and J. Tavantzis, *On the asymptotic behavior of solutions of certain quasilinear parabolic equations*, J. Diff. Eq. 54 (1984), 97-120.

[OPP] A.K. Oppenheim and L. Zajac, *Dynamics of an explosive reaction center*, AIAA J. 9 (1971), 545-553.

[PAO1] C.V. Pao, *Successive approximations of some nonlinear initial-boundary value problems*, SIAM J. Math. Anal. 5 (1974), 91-102.

[PAO2] C.V. Pao, *Positive solutions of a nonlinear boundary value problem of parabolic type*, J. Diff. Eq. 22 (1976), 145-163.

[PAO3] C.V. Pao, *Asymptotic behavior and nonexistence of global solutions for a class of nonlinear boundary value problems of parabolic type*, J. Math. Anal. Appl. 65 (1978), 616-637.

[PAO4] C.V. Pao, *On nonlinear reaction-diffusion systems*, J. Math. Anal. Appl. 87 (1982), 165-198.

[PEL] L.A. Peletier, *Asymptotic behavior of the solutions of the porous media equation*, SIAM J. Appl. Math. 21 (1971), 542-551.

[PRO] G. Prodi, *Teoremi di esistenza per equazioni alle derivate parziali non lineari di tipo parabolico*, Rend. Ist. Lombardo 86 (1953), 3-47.

[PRT] M.H. Protter and H.F. Weinberger, *Maximum Principles in Differential Equations*, Prentice-Hall, New Jersey, 1967.

[PUE] J. Puel, *Existence comportement à l'infini et stabilité dans certaines problémes quasilinéares elliptiques et paraboliques d'ordre 2*, Ann. Scuola Norm. Sup. Pisa Cl. Sci. (4) 3 (1976), 89-119.

[RED] R. Redheffer and W. Walter, *Invariant sets for systems of partial differential equations*, Arch. Rat. Mech. Anal. 67 (1977), 41-52.

[SAT1] D. Sattinger, *Monotone methods in nonlinear elliptic and parabolic boundary value problems*, Indiana Univ. Math. J. 21 (1972), 979-1000.

[SAT2] D. Sattinger, *Topics in Stability and Bifurcation Theory*, Lecture Notes in Mathematics 309, Springer-Verlag, New York, 1973.

[SCH] K. Schmitt, *Boundary value problems for quasilinear second order elliptic equations*, Nonlinear Analysis 2 (1978), 263-309.

[SCU] V. Schuchmann, *About uniqueness for nonlinear boundary value problems*, Math. Ann. 267 (1984), 537-542.

[SER] J. Serrin, *A symmetry problem in potential theory*, Arch. Rat. Mech. Anal. 43 (1971), 304-318.

[SHI] R. Shivaji, *Remarks on an s-shaped bifurcation curve*, J. Math. Anal. Appl. 11 (1985), 374-387.

[SMO] J. Smoller, *Shock Waves and Reaction-Diffusion Equations*, Springer- Verlag, New York, 1983.

[SPE] R. Sperb, *Maximum principles and their applications*, Academic Press, New York, 1981.

[STE] D.S. Stewart, *Shock initiation of homogeneous and heterogeneous condensed-phase explosives with a sensitive rate*, Comb. Sci. Tech. 48 (1986), 309-330.

[STR] R.A. Strehlow, *Fundamentals of Combustion*, Krieger Publishing, New York, 1979.

[TAL1] P. Talaga, *The Hukuhara-Kneser property for parabolic systems with nonlinear boundary conditions*, J. Math. Anal. Appl. 79 (1981), 461-488.

[TAL2] P. Talaga, *The Hukuhara-Kneser property for quasilinear parabolic equations*, Nonlinear Analysis: TMA 12 (1988), 231-245.

[TRO1] W. Troy, *Symmetry properties in systems of semilinear elliptic equations*, J. Diff. Eq. 42 (1981), 400-413.

[TRO2] W. Troy, *The existence of bounded solutions for a semilinear heat equation*, SIAM J. Math. Anal. 18 (1987), 332-336.

[WAL] W. Walter, *Differential and Integral Inequalities*, Springer, New York, 1970.

[WEI] H. Weinberger, *Invariant sets for weakly coupled parabolic and elliptic systems*, Rend. Mat. 8 (1975), 295-310.

[WES1] F.B. Weissler, *Semilinear evolution equations in Banach spaces*, J. Funct. Anal. 32 (1979), 277-296.

[WES2] F.B. Weissler, *Existence and nonexistence of global solutions for a semilinear heat equation*, Israel J. Math. 38 (1981), 29-40.

[WES3] F.B. Weissler, *Single point blowup of semilinear initial value problems*, J. Diff. Eq. 55 (1984), 204-224.

[WES4] F.B. Weissler, *An L^∞ blowup estimate for a nonlinear heat equation*, Comm. Pure Appl. Math. 38 (1985), 291-296.

[WES5] F.B. Weissler, *L^p-energy and blowup for a semilinear heat equation*, Proc. Symp. Pure Math. 45, Part 2, (1986) 545-552.

[WST] H. Westphal, *Zur Abschätzung der Lösungen nichtlinear parabolischer Differentialgleichungen*, Math. Z. 51 (1949), 690-695.

[WHI] G.B. Whitman, *Linear and Nonlinear Waves*, Wiley, New York, 1974.

[WIL1] F.A. Williams, *Combustion Theory*, Addison-Wesley, Massachusetts, 1969.

[WIL2] F.A. Williams, *Theory of combustion in laminar flows*, Ann. Rev. Fluid Mech. 3 (1971), 171-188.

[YAN] Z.Q. Yan, *Invariant sets and the Hukuhara-Kneser property for parabolic systems*, Chin. Ann. Math. 5B (1) (1984), 119-131.

[YOS] K. Yosida, *Functional Analysis*, Springer-Verlag, New York, 1980.

[ZEL] Y.B. Zeldovich, Zh. Eksperim i. Teor. Fiz. 9 (1939), 12.

[ZLN] T.I. Zelenyak, *Stabilization of solutions of boundary value problems for a second-order parabolic equation with one space variable*, Differential Equations 4 (1968), 17-22.

Index

a *priori* bounds 47, 72, 133, 137, 139, 151, 157, 160, 161
activation energy 2, 7, 9, 10, 13
  asymptotics 7, 8
Arrhenius law 2
Arzela-Ascoli Theorem 83

bifurcation curve 15, 36, 38, 39, 41, 45
blowup 47, 48, 53, 54, 64, 82, 86, 87, 107, 127, 128, 136, 161
  everywhere 107, 120, 121, 124, 127
  point 47, 65, 66, 68, 69, 120, 121, 129
  single-point 65, 66, 87, 107, 120, 122, 123, 124
  singularity 135, 136 (*see also* hot spot)
  time 47, 48, 53, 54, 55, 56, 57, 58, 60, 63, 64, 69, 74, 87, 107, 118, 120, 126, 127, 129, 135, 161
boundary conditions
  Dirichlet 15, 95, 104, 106, 129
  mixed 60, 103, 104
  Neumann 95

cap 23
  maximal 25
  optimal 25
Cauchy's inequality 145, 148, 149, 151, 152, 153, 154
chemical reaction
  exothermic 1
  one-step 2
coefficient
  diffusion 2
  stoichiometric 2, 6
  thermal conductivity 4
  thermal diffusivity 7
  viscosity 3
comparison 47, 48, 54, 75, 88, 90, 92, 93, 106, 107, 113, 157
concentration 10
conservation equations
  complete system of 1, 5, 129
  energy 3, 10, 11
  mass 1, 11
  momentum 3, 10
  species 2, 10

density bounds 140
Dini's Theorem 49

eigenvalue condition 95, 97, 100, 101, 102, 103
energy
  density 140, 146, 147
  estimates 130, 140, 160, 161
  integral 47, 83
  internal 4
  kinetic 4
  total 141
enthalpy 4
Euler coordinates 12, 137

Fick's law 2, 4
final time analysis 69, 71
first eigenvalue 17, 18, 19, 20, 54, 63
first variational problem 57, 59

flux
- condition (strong) 96, 97, 98, 100, 101
- condition (weak) 96, 98, 101, 102, 103
- diffusion 4
- energy 4

Frank-Kamenetski parameter 8, 20, 134

gas
- one-dimensional model 129, 136
- parameter 7
- universal constant 3

Gelfand problem 15, 17, 18, 20, 33, 39, 43, 44, 45, 46
- perturbed 15, 20, 38, 43, 46

gradient system 129
Green's identity 40, 41, 54, 58, 108, 109, 110
Green's function 50
Gronwall's inequality 144, 154

heat
- of formation 4
- of reaction 7
- release 134
- specific 4

Hölder continuity 86, 99, 155, 156
locally 48, 50
Hölder inequality 143, 145, 146, 155, 158
Hopf lemma 22, 28, 98
hot spot 13, 87, 128, 161 (*see also* blowup singularity)

ignition model 7, 13, 14, 86
- gaseous fuel 10, 12
- gaseous reactive-diffusive 11, 14, 107, 127
- nondiffusive 11, 107, 125, 128
- reactive Euler 12, 14, 129, 133
- solid fuel 8, 11, 47, 64, 88, 92, 107, 127

ignition (induction) period 1, 9, 12, 161
Implicit Function Theorem 75
invariance 13, 47, 48, 88, 94, 95, 97, 98, 99, 101, 103, 104, 106, 111, 112, 115, 117, 119, 160
Inverse Function Theorem 35
isoperimetric inequality 17

Jensen's inequality 54, 57, 58, 59, 63

Lagrange variable 137
Lebesgue Dominated Convergence Theorem 50
Leray-Schauder degree theory 88, 99, 101, 102
Lewis number 7
limit cycle 36
Lipschitz continuous 67, 68, 91, 93, 115, 116, 139, 140, 161
- global 85, 114, 138, 147, 153
- locally 22, 49, 75, 89
- uniform 83, 90, 91, 102

lower solution 15, 16, 17, 18, 19, 20, 46, 47, 48, 49, 50, 53, 86, 106, 112
Lumer-Phillips Theorem 109

majorant function 93
maximal solution 21, 39
maximum principle 15, 17, 19, 22, 23, 28, 29, 31, 33, 43, 61, 62, 63, 64, 65, 66, 68, 69, 71, 74, 75, 76, 87, 88, 91, 105, 116, 122, 123, 125, 140, 157
Mean Value Theorem 28, 29, 30
method of lines 55
method of moving parallel planes 15, 22, 23
minimal solution 21, 39, 44, 49, 50, 52, 53, 63

minorant function 93
model (*see also* ignition model)
  nondimensional 6, 7
  one-dimensional gas 129, 136
  small fuel loss 8, 46
  steady-state 8, 47, 62, 63, 76
multiplicity 15, 21, 33, 38, 46

node
  spiral 36
  unstable 36
nonlinear eigenvalue problem 16

orbit
  heteroclinic 34,36,38,42
  periodic 36

parabolic quasilinear system 1, 95
Perron method 86
Prandtl number 7

quasimonotone function 46, 88, 90, 91, 93, 106, 113

radial symmetry 15, 21, 22, 29, 30, 31, 33, 43, 46, 47, 64, 67, 71, 83, 87, 112, 115, 120, 124
reaction-diffusion equations 7, 8
replacement vector 90
Schauder's interior estimates 83
self-similarity 47, 48, 72, 74, 82, 87, 135
semigroup theory 106, 107, 128
  adjoint 109
  analytic semigroup 108, 109
  closed,convex cone 112
  contraction semigroup 109, 116
  dissipative 109, 114
  equicontinuous semigroup 108
  infinitesimal generator 108
  Lumer-Phillips Theorem 109
solution
  bell-shaped 39, 44, 45
  generalized 138, 139
  large-small 38, 45
  maximal 21, 39
  minimal 21, 39, 44, 49, 50, 52, 53, 63
  profile 38, 39, 47, 87, 129
  singular 73, 74, 86
  small 38
  small-small 38, 45
  steady-state 48, 73
space-time parabola 47, 87
spectrum 16, 57, 60
symmetrization 17, 21

temperature bounds 157
temperature perturbation 10
tensor
  deformation 3
  identity 3
  stress 4
thermal event
  subcritical 9, 119
  supercritical 9, 53, 86
thermal runaway (*see* blowup)
time
  acoustic 7, 9
  conduction 7, 9, 10, 11
  reference 7, 10

upper solution 15, 16, 17, 18, 19, 20, 46, 47, 48, 49, 50, 54, 62, 86, 106, 117, 120

velocity bounds 152
velocity perturbation 12
viscosity 129, 136, 161

Young's inequality 146, 158

Applied Mathematical Sciences

---

55. *Yosida:* Operational Calculus: A Theory of Hyperfunctions.
56. *Chang/Howes:* Nonlinear Singular Perturbation Phenomena: Theory and Applications.
57. *Reinhardt:* Analysis of Approximation Methods for Differential and Integral Equations.
58. *Dwoyer/Hussaini/Voigt (eds.):* Theoretical Approaches to Turbulence.
59. *Sanders/Verhulst:* Averaging Methods in Nonlinear Dynamical Systems.
60. *Ghil/Childress:* Topics in Geophysical Dynamics: Atmospheric Dynamics, Dynamo Theory and Climate Dynamics.
61. *Sattinger/Weaver:* Lie Groups and Algebras with Applications to Physics, Geometry, and Mechanics.
62. *LaSalle:* The Stability and Control of Discrete Processes.
63. *Grasman:* Asymptotic Methods of Relaxation Oscillations and Applications.
64. *Hsu:* Cell-to-Cell Mapping: A Method of Global Analysis for Nonlinear Systems.
65. *Rand/Armbruster:* Perturbation Methods, Bifurcation Theory and Computer Algebra.
66. *Hlaváček/Haslinger/Nečas/Lovíšek:* Solution of Variational Inequalities in Mechanics.
67. *Cercignani:* The Boltzmann Equation and Its Applications.
68. *Temam:* Infinite Dimensional Dynamical System in Mechanics and Physics.
69. *Golubitsky/Stewart/Schaeffer:* Singularities and Groups in Bifurcation Theory, Vol. II.
70. *Constantin/Foias/Nicolaenko/Temam:* Integral Manifolds and Inertial Manifolds for Dissipative Partial Differential Equations.
71. *Catlin:* Estimation, Control, and the Discrete Kalman Filter.
72. *Lochak/Meunier:* Multiphase Averaging for Classical Systems.
73. *Wiggins:* Global Bifurcations and Chaos.
74. *Mawhin/Willem:* Critical Point Theory and Hamiltonian Systems.
75. *Abraham/Marsden/Ratiu:* Manifolds, Tensor Analysis, and Applications, 2nd ed.
76. *Lagerstrom:* Matched Asymptotic Expansions: Ideas and Techniques.
77. *Aldous:* Probability Approximations via the Poisson Clumping Heuristic.
78. *Dacorogna:* Direct Methods in the Calculus of Variations.
79. *Hernández-Lerma:* Adaptive Markov Control Processes.
80. *Lawden:* Elliptic Functions and Applications.
81. *Bluman/Kumei:* Symmetries and Differential Equations.
82. *Kress:* Linear Integral Equations.
83. *Bebernes/Eberly:* Mathematical Problems from Combustion Theory.